America Project Making of, Émile Saigey, Thomas Freeman Moses

The Unity of Natural Phenomena

A Popular Introduction to the Study of the Forces of Nature

America Project Making of, Émile Saigey, Thomas Freeman Moses

The Unity of Natural Phenomena
A Popular Introduction to the Study of the Forces of Nature

ISBN/EAN: 9783744690379

Printed in Europe, USA, Canada, Australia, Japan

Cover: Foto ©berggeist007 / pixelio.de

More available books at **www.hansebooks.com**

Science for the People.

No. I.

THE UNITY OF NATURAL PHENOMENA.

By Emile Saigey.

IN PRESS:

THE SCIENCE AND PHENOMENA OF HEAT.

From the French of Amedee Guillemin.

THE PHENOMENA OF PLANT LIFE AND SEXUALITY OF NATURE.

By L. Hartley Grindon.

THE PHENOMENA OF SOUND.

OF

NATURAL PHENOMENA.

A POPULAR INTRODUCTION TO THE STUDY OF

THE FORCES OF NATURE.

FROM THE FRENCH OF

M. EMILE SAIGEY.

WITH

AN INTRODUCTION AND NOTES

BY

THOMAS FREEMAN MOSES, A. M., M. D.,

PROFESSOR OF NATURAL SCIENCE IN
URBANA UNIVERSITY.

BOSTON:
ESTES AND LAURIAT.
1873.

CONTENTS.

CHAPTER IV.

CHAPTER V.

CHAPTER VI.

CHAPTER VII.

TRANSLATOR'S INTRODUCTION.

Since the discovery of the laws of gravity, more than two hundred years ago, no scientific achievement has been so fruitful in results as the determination of the mechanical equivalent of heat. In 1842 Dr. Mayer, of Heilbronn, in Swabia, demonstrated that the blow of a hammer weighing four hundred and twenty-four kilograms upon an anvil, with the velocity it has acquired by falling through a distance of one metre, produces an elevation of temperature equal to one degree centigrade. This experiment has wrought a marvellous change in all our conceptions of Matter and Force, and still gives promise of results which the imagination fails to grasp. It is now considered as demonstrated that heat, electricity, light, magnetism, chemical attraction, muscular energy, and mechanical work, all are but exhibitions of one and the same power acting through matter. The molecules of matter, variously stirred by this all-pervading force, are thrown into waves, which strike against our senses, and the motion thus communicated to our nerves impresses us as heat, sound, or light, according to the rapidity and breadth of the undulations. If an undulation of one sort is interfered with, another immediately succeeds of exactly the same strength. Light ransforms itself into chemical action, this into heat, and heat into motion. Finally, all these modes of motion are not only mutually convertible, but they may also be turned into *mechanical work.* The amount of work which a fixed quantity of each can do is termed its "*mechanical* equivalent." In other words, *force* is a constant energy, never increasing or diminishing in absolute value. It is like a stream, now flowing through uninterrupted channels, silently and imperceptibly, now thrown into gentle undulations as it comes into contact with the subtle forms of matter, and now giving striking displays of its power as it meets with greater resistance. Light, sound, heat, the invisible flow of the terrestrial magnetic currents, as well as the aurora and zodiacal light, the flow of the sap in plants, and the circulation of the blood in animals, are all exhibitions of this single Force.

From the unity of Force the induction was easy and natural to the unity of Matter. The whole tendency of modern physical science is to do away with the notion of an indefinite number of primary elements, and to substitute instead a smaller and smaller number of primitive forms of matter. We must admit, however, that this notion is not strictly a modern one, for the idea that matter is identical, and only its forms various, is as old as Aristotle. The results of experimental investigation at first, and for a long time, cast discredit upon this ancient doctrine, but modern chemical and physical research serve only to confirm it. Thus a great cycle of human thought is completed.

What, in brief, is the theory of modern science in regard to matter?

By changes in the mode of aggregation of the atoms, which changes depend chiefly upon the degree and kind of motion with which they are endowed, matter appears to us under certain definite forms, as water, air, iron, etc. The analysis of many substances, once supposed to be compound, into simple ones, leads naturally to the conclusion that such an analysis is possible in the case of many substances now placed in the category of elements. That wonderful class of elements, the metals, would appear to be the most stable of all; but who can foresee the result of spectroscopic investigation upon these, even? That there exists among them some common elemental principle, of which each metal is a variety, is an idea already broached. A comparison of their characteristic spectrum lines may afford startling results, surpassing the dream of the alchemists. Recent applications of spectrum analysis to the nebulæ and fixed stars, in connection with the theory of the evolution of suns and planets, in accordance with the nebular hypothesis, are strikingly suggestive. The spectra of some nebulæ give evidence of but two elements, nitrogen and hydrogen, and a classification of the fixed stars and nebulæ may be so arranged as to exhibit a gradual increase in complexity of structure.* In a word, variety of form and unity of substance, the evolution of the complex from the simple, of the heterogeneous from the homogeneous, are the fundamental principles of modern philosophy.

The author of the following Essay assumes that there is but one material substance, and that this substance is the Ether. This ether is the primitive and most subtle element; it is the tissue out of which the entire universe is wrought — a kind of mineral protoplasm, in fact. This theory is clearly and compactly stated in the Author's General Hypothesis. The subject of the ether is engrossing so much attention at

* See an interesting article in Popular Science Monthly, January, 1873, entitled Evolution and the Spectroscope..

the present time, it may not be out of place here to mention some of the views entertained in regard to it by former and contemporary writers.

The revival of the doctrine of the ether, and its support by many of the most eminent physicists of the present day, is a remarkable event in the history of science. There is indeed nothing new under the sun. The ether dates back to the early days of scientific research, and first occurs as a subdivision of the *air*, one of the four elements of Empedocles. It is described as filling all the interplanetary spaces, and also as penetrating and enveloping the particles composing the heavenly bodies, impregnating them with certain effluvia and influences. The Greek poet Orpheus hails it as "the bright, life-giving element" flowing around the sun, moon, and stars, and calls it the *blastema*, or universal germ of things. In 1664 the undulatory theory of light demanded the hypothesis of a homogeneous medium for the propagation of the light waves. This theory received its first definite statement from Huyghens, in 1690. Light was judged to be the vibration of an ether. Even Newton, while holding to his emission theory of light, acknowledged the probability of the existence of such a medium,* and speaks of the universal ether as the

* Whewell, History Inductive Sciences, vol. ii. p. 89.

"sensorium of God." Newton's authority in science served to hold the new doctrine of light in abeyance for more than a hundred years; it was not until the beginning of the nineteenth century that it was placed beyond a doubt by the labors of Dr. Thomas Young. Near the middle of the eighteenth century, while the Newtonian theory was yet in the ascendency, Swedenborg published his Principia, in which he maintains the existence of an ether as one of five elementary forms of matter. These are the solar vortex, which is the cause of gravity; second, the magnetic element; third, the ether; fourth, the atmosphere; and lastly, the aqueous vapor. The ether is the third of these elements in the order of succession, proceeding from the sun to the planets. Its vibrations and undulations are the cause of light and heat. Motion, diffused from a centre through a contiguous medium of ether particles, produced light and also heat. Our earth is surrounded with an ethereal aura, the position of which is intermediate between the solar vortex and our own atmosphere, thus serving to keep up the contiguity of expanse between the sun and our earth. It is in the state of ether that matter has become "sufficiently gross to be perceptible to our senses by its effects." Swedenborg, in the calm and philosophic spirit characteristic of him, took no pains to enforce these doctrines upon the attention of his own age. This indifference was shared by his contemporaries at a time when the scientific world was distracted by the conflicting theories of light. In 1671 Leibnitz published a new physical hypothesis, in which he deduced the causes of most physical phenomena from a single universal motion. The particles of the earth are endowed with separate motions, which give rise to shocks, from which results an agitation of the ether radiating in all directions.

Passing down to our own times, we find various theories in regard to the ether. That of Hartman is in substance as follows: There are two kinds of atoms, ethereal and corporeal; bodies are made of spherically-shaped atoms, which act equally in all directions, the activity proceeding from the centre. Between the molecules of these bodies are scattered the atoms of the ether, and they also fill interstellar space. The corporeal atoms have a tendency to converge towards a single point, but are kept apart by the ether atoms. Between the corporeal and the ethereal atoms a mutual repulsion exists. Another writer, Spiller,* calls the "world-ether the soul of the universe." By such an eminent authority as Tyndall, the ether is recognized

* See Article on Philip Spiller's Aetherism, in the New York Evening Post, January 24, 1873.

as the interstellar medium filling all space. "The luminiferous ether fills stellar space, makes the universe a whole, and renders the intercommunication of light and energy between star and star possible. But the subtle substance penetrates farther; it surrounds the very atoms of solar and liquid substances." And again, "The intellect knows no difference between great and small; it is just as easy to conceive of a vibrating atom as of a vibrating cannon-ball; and there is no more difficulty in conceiving of this Ether, as it is called, which fills space, than in imagining all space to be filled with jelly. You must imagine the atoms vibrating, and their vibrations you must figure as communicated to the ether in which they swing, being propagated through it in waves; these waves enter the pupil, cross the ball of the eye, and break upon the retina at the back part of the eye. The act, remember, is as real and as truly mechanical as the breaking of the sea-waves upon the shore. Their motions are communicated to the retina, transmitted thence along the optic nerve to the brain, and there announce themselves to consciousness as light."

The essential difference between the foregoing theories and that of M. Saigey may be stated thus: In the former the ether is regarded as filling all the spaces between the stars and planets and the atoms of bodies.

M. Saigey believes it to be all this and more. He makes it also the *constitutive* element of the atoms themselves. "The atom and motion, behold the universe!" Such is his enthusiastic language. Ether in a state of motion fills all space. The ethereal atoms form societies, which are the molecules of bodies. A body is a collection of these societies of molecules. Between these atoms, molecules, and bodies, exchanges of motion take place, which constitute heat, light, chemical affinity, and gravity. These exchanges depend upon the relations of mass and velocity.

Granting these premises, the remaining point in M. Saigey's hypothesis — that the laws which govern the interaction of this primitive force and matter are none other than the laws of mechanics — is but a logical deduction. These laws depend upon geometrical relations, and may be mathematically expressed. But it must be confessed that the ether itself, as he defines it, is a hypothetical substance, and its existence lacks "mechanical confirmation." Still, this is not an insuperable objection, for all scientific induction must start with, and be based upon, hypothesis. The ether is a scientific necessity. To illustrate: In the mental conception which we form of the relations of bodies in space, and of the parts of a body among themselves, which relations we symbolize in geometrical figures, we are obliged to have recourse to a hypothetical point. In geometry solids are regarded as generated by the motion of surfaces, surfaces by the motion of lines, and lines of points. Beyond this we are not able to go. By an analogous train of reasoning, M. Saigey builds up the universe out of the ethereal atom by the aid of motion. Masses are made of compound particles; the particles are aggregations of molecules, and the molecules may be resolved into atoms. Behind this veil of atoms lies the Infinite. Matter is a series of orderly changes from the *immaterial*, becoming more and more gross until recognized by the senses.* Matter is, "at bottom, essentially mystical and transcendental." †

From the organic forms of matter M. Saigey passes to living beings, to plants and animals; and these are likewise brought into his hypothesis. The primitive cell of the plant, as well as the embryonic germ of the animal, are formed out of the materials of the inorganic world. In both a series of motions succeeds each other, according to a fixed order. This series of motions possesses a special character, it is true; but their transformations obey the laws of molecular mechanics. The origin of force, the nature of life, human person-

* Swedenborg, Principia, vol. i. chap. ii.

† Fragments of Science, p. 415.

ality — questions naturally suggested in this connection — are considered by M. Saigey as entirely foreign to his argument, and are, therefore, left untouched. All scientific men have not shared the forbearance of our author. Emotions and ideas display outward phenomena which are subject to the laws of thermodynamics. Without question these *phenomena* are legitimate subject of scientific inquiry. Some time ago a series of experiments was published, showing that the operations of the mental and emotional faculties are accompanied by a change of temperature in the brain. The greatness of an idea and the strength of an emotion may peradventure be measured, and their mechanical equivalent determined. Is it a logical inference from this that mind is a *quality* of nerve tissue? and does science lay us under any obligations to accept the dogmas of a school of philosophy which teaches that growth, development, and human progress are results of self-determining processes, inherent in the very nature of things? There is, as there should be, a spirit of fearless inquiry abroad, which will not stop at prescribed bounds. Unhappily it is accompanied by a spirit of irreverence, for which a Newton or a Kepler would have blushed. The scientific method of arriving at truth may be the most exact and satisfactory one, but it is not the only one. The fundamental facts of revelation, confirmed by the methods of a spiritual science, rest upon proofs as sure and convincing, to say the least, as any established by natural science. Between these two methods the utmost harmony may and ought to exist. In a recent article in the Contemporary Review, Dr. Carpenter expresses his opinion that science *points to* the origination of all power in *Mind;* and, further, that there are satisfactory grounds for the belief that the phenomena of the material universe are the expressions of a Mind and Will, of which man's is the finite prototype. To admit this will be to admit the existence of the supernatural, always working in and through the natural. Force ceases to be a blind attribute of matter, and becomes a living, active principle, spiritual in its character.

I cannot forbear quoting here the sublime language of one who may rightfully be called the greatest of living naturalists, one who, as there is recent ground for believing, has thus far withstood the current of the Darwinian theories with a firmness and stability like that of his native Alps.

"The combination in time and space of all these thoughtful conceptions (just recapitulated), exhibits not only thought, it shows premeditation, power, wisdom, greatness, prescience, omniscience, providence.

In one word, all these facts in their natural connection proclaim aloud the one God, whom man may know, adore, and love; and Natural History must, in good time, become the analysis of the thoughts of the Creator of the universe, as manifested in the animal and vegetable kingdoms." *

THOS. FREEMAN MOSES.

URBANA UNIVERSITY, March 2, 1873.

* Agassiz, Essay on Classification, p. 135.

AUTHOR'S INTRODUCTION.

THAT heat and mechanical force are mutually equivalent is a fact familiar to all who have interested themselves in the progress of science. Everywhere we see heat converting itself into force, and force into heat. In the steam engine, for example, the heat disengaged by the combustion of the coal is turned into the labor produced by the shaft of the engine. On the other hand, if you turn a wheel in a body of water, the water becomes warm; if you rub together two blocks of ice, the ice melts. All around us, in the work of every-day life, we see a certain quantity of heat disappearing at the same time that a certain amount of force is produced; and the converse is equally well known from the most familiar facts. Simple as this notion appears to us, now that it has become a part of our current ideas, its discovery is, nevertheless, the principal achievement of modern science.

The works of Mr. Joule, the eminent philosopher of Manchester, those of Mr. Jules-Robert Meyer, of Heilbronn, of Hirn, the Colmar engineer, after having determined the conditions of the convertibility existing between heat and mechanical power, have fully brought to light the principle itself, as well as the reason of this convertibility. By force is meant the displacement of a body. Now, heat, as every one will at this day admit, is a molecular movement, a displacement of molecules: is it not perfectly natural, then, that these two phenomena should replace each other, according to a fixed relation, and that between these two kinds of motion there should exist a ready convertibility governed by the common laws of mechanics?

From the moment of the introduction of this precise and well-defined idea into science, every branch of physics has undergone, in some degree, a renovation. Upon many questions the new theory has thrown a direct light; upon others it has furnished many luminous hints, and been the incentive to useful research. Around undisputed facts brought to light by the study of heat, other facts, less well established, have ranged themselves, then ingenious conjectures, and from this impulse of ideas has sprung up a new conception of nature, which has already engaged the attention of many scientific minds.

In turning our attention to this new way of viewing natural phenomena, we find it, at the outset, a difficult one to define.

The unity of physical forces, — such is the general formula embracing the various considerations, of which we shall attempt a rapid review.

In the system before us all the forces of nature are traceable to the same principle, and are mutually convertible under certain fixed laws, which are none other than the laws of mechanics. Such is a rude and general statement of the new theory, a statement accepted by different scientific minds, not without certain restrictions. Those even who are almost agreed as to the principle itself differ when it becomes necessary to deduce certain consequences on the subject of the condition of matter and the constitution of the globe. It is here that we meet with our first difficulty. We shall not presume to set forth, in subjects so important, a collection of opinions which are merely personal; nor, on the other hand, may we venture to say, that even among the partisans of this new theory has there been harmony sufficient to establish a true scientific system.

In undertaking the study of the equivalence of heat

and motion we readily find some useful guides. Complete treatises upon this subject already exist, and he who wishes to examine its leading points in a precise and substantial form may refer to the excellent lectures of Mr. Verdet, published in the Memoirs of the Chemical Society of Paris, under the title of *Exposé de la Théorie Méchanique de la Chaleur.* In examining the unity of the physical forces, however, we have no such aid as the above. It would, therefore, seem opportune at this time to give some definite outline to ideas which have hitherto remained vague and ill defined. We have already made an attempt in this direction in a number of articles recently published in the *Revue des deux Mondes,* and which form the substance of the Essay now presented to the public. It will be gratifying to us if our attempt shall call forth new light upon the correlation of natural phenomena, and if our Essay shall hasten the publication of some important work on this subject.

In 1864, Father Secchi, director of the observatory of the Roman College, published an interesting volume, *L'Unità delle forze fisiche, saggio di filosofia naturale.* Father Secchi cordially adopted the idea that the physical forces are all traceable to one and the same principle. The study of astronomical phenomena has itself furnished the grounds for this opinion. In reflecting upon the force of gravity, which gives motion to the heavenly bodies, he was not accustomed to regard it as an elementary principle, but traced it back to a still more general law, of which this force is only a consequence. His work contains novel and original suggestions upon this subjeet. At the same time this book has the character of a compendium of physical science. It reviews in a summary manner the facts which to-day compose the resources of science; only accidentally and at intervals does he touch upon the general principles which these facts suggest. We do not find here any theory fully set forth, by which the forces of nature may be traced back to their unity.

We might mention a still older work, that of M. de Boucheporn, published in 1853, under the title of *Principe général de la Philosophie naturelle.* It is a work compiled with care and zeal — one of those books in which a man condenses the thought of a lifetime. M. de Boucheporn seizes with boldness upon the synthesis of natural phenomena. He recoils before no difficulties, he meets face to face every obstacle. Herein lies the merit, as well as the defect, of his work. M. de Boucheporn trusts himself too hastily and too entirely to what are mere guesses. It is wonderful to observe how a conjecture becomes a certainty with him as soon as he can make use of it in accounting for certain facts. It is equally a wonder to see how supple facts become in his hands, and how readily they adapt themselves to the demonstrations which he exacts of them. Let us add that at the time when M. de Boucheporn published the *Principe général de la Philosophie naturelle*, the new theory of heat had not assumed a definite position in science; it was but just being evolved, and its results were but poorly comprehended, and the author, while he did not ignore them, yet derived but little benefit therefrom. His book, too, while it remains full of interest in what relates to astronomy, has lost much of its value in the part treating of the laws of physics, properly so called. At the same time the moment we set out from the facts revealed by the study of heat, the general theory which we wish to develop can hardly be presented as otherwise than hypothesis. As we just now remarked, a serious difficulty presents itself when we try to bring under a precise definition that new conception of nature to which recent investigations have given rise. In what language shall we present it in order that it may not appear rash to some, to others chimerical, and to many useless? Within what bounds must we retain it, that we may not seem to overstep the facts? May we be permitted to adopt the following plan, not indeed as the wisest, but as likely to throw the most light upon the subject?

We shall begin by setting forth in its entirety, and in all its simplicity, this grand hypothesis which we have designated under the name of the *Unity of Physical Forces*, and we shall try to demonstrate its immediate consequences. We shall not occupy ourselves at the outset with bringing forward proofs to establish this opinion.

It is only, in conclusion, that we shall endeavor to indicate on what grounds the hypothesis rests, and the restrictions and modifications will gradually present themselves. In this presentation of proofs it can easily be seen how much belongs to experience, how much to imagination; what can be believed without scruple, and what must be the subject of doubt until further information.

With this understanding at the outset, we beg leave to sketch our hypothesis in all its force, without being obliged to weaken it, as we proceed, with restrictive suggestions.

THE

UNITY OF NATURAL PHENOMENA.

CHAPTER I.

THE GENERAL HYPOTHESIS.

I.

Atoms and Motion.—Physical Phenomena may be reduced to a single Principle, and considered as the Effects of Motion.

THAT matter exists throughout the universe in an unvarying quantity is now an undisputed fact, and one beyond the reach of controversy. Never is it created anew, never destroyed; it simply passes through transformations. The progress made by chemistry in the beginning of this century set forth this truth in all its clearness, and made it, in a manner, palpable.

What are, in fact, the properties of matter? First, impenetrability, as implied in its definition, a portion of matter being that which occupies a share of space to the exclusion of any other portion ; second, inertia, it being the principal result of human experiment, and the foundation, indeed, of mechanics, that matter is set in motion only when it receives an impression, and loses its motion only in communicating it.

We can say the same of motion that we just now said of matter — it is neither created nor destroyed ; its quantity is invariable. With motion, as with matter, it is only a question of transformation.

Here the idea of Force claims our attention. What is force in the language of physics and mechanics? It is a cause of motion ; but what idea does this convey to us? The cause of a motion is another motion. We will dispense, then, if it is possible, with this idea of force, or rather, — for it is necessary to employ customary terms in order to be intelligible, — we will understand by force whatever causes one motion to give place to another motion.

If now, leaving these abstract considerations and entering the domain of facts, we ask what are the physical phenomena which appeal to our senses, heat, light, electricity, magnetism, we find it demonstrated that heat is one kind of motion, and that light is another, and we are made to perceive that it is the same with electricity and magnetism. There is nothing, then, which need surprise us if one of these motions should engender the other; that heat should be transformed into electricity, and electricity into light. When the solar rays draw up water from the surface of rivers and lakes, when clouds are formed, when these clouds become charged with electricity and the lightning flashes, and when the watery vapor falls in rain to the ground, we see under these various appearances only a series of successive motions. Not only do we find, at the end of the phenomenon, the whole quantity of water employed in it, but the mind follows easily its various modifications from its first motion. Now, it will be understood that these transformations take place according to certain fixed relations. If the various kinds of motion be measured according to certain established units, all these units are reducible to a common scale; a unit of heat corresponds always to 425 kilogrammetres,* to 425 units of mechanical power; there is an analogous relation between the electric unit and the heat unit, and so on.

If, now, we touch upon another order of facts, if we consider another group of forces, the cohesion which

* The kilogrammetre is the power represented by a kilogramme (2¼ pounds nearly) elevated to the height of a metre (about 40 inches).

keeps bodies either in a solid or liquid state, the chemical affinity which attracts molecules of different kinds, that force of gravity, in a word, by virtue of which bodies tend to move towards each other, the new theory enables us to perceive still farther how the play of all these forces is reducible to certain communications of motion. Here is, for example, a piece of lead, whose molecules adhere in such a manner as to form a solid block. I know that in heating them, in communicating to them a certain kind of motion, I shall destroy the cohesion by virtue of which the block remains solid, and I shall subject it to a different cohesion, which belongs to the liquid state. Heating it still more, and thus augmenting the amount of communicated motion, I shall destroy even this kind of cohesion, and reduce the metal to vapor. Hence may we not suppose that the cohesion which held together the molecules of the lead was a motion relatively of those molecules? Whatever we destroy by a motion must be itself a motion. Cohesion, we say, proceeds from a relative motion. Do we not see it in some cases result simply from a common velocity imparted to neighboring molecules.

When a jet of water, for example, escapes from an orifice under strong pressure, does it not assume a solid form, and is there not a kind of cohesion resulting from the fact that all its molecules in the same plane move onward with an equal velocity? The familiar example we have cited will not be regarded as a demonstration of facts. At present we do not endeavor to crowd together phenomena; we shall strive only to show in what manner the new theory presents itself to our view. We would not discuss its revelations. We seek only to present them to view.

In regard to chemical affinity, we shall only be allowed to say a word here, for in many respects its action is similar to that of cohesion, and somewhat analogous to that of gravity. When, under certain conditions, particles of oxygen and carbon meet, they are precipitated upon each other after the manner of heavy bodies, and when they are combined to form the oxide of carbon or carbonic acid, the fixed state which they enter upon may be compared to that of planetary bodies which revolve about each other.

What, then, is gravity? What is that mysterious force which causes two bodies to attract each other in the direct ratio of their masses and the inverse ratio of their distance? Two bodies attract each other! Then matter is not inert! Would there not appear to be a real contradiction between the two terms, matter and inertia?

The question is one well worthy our close attention and examination.

Here are two molecules of matter. Is it a sound notion to imagine them as setting out voluntarily from a state of repose for the purpose of approaching each other again? Strictly speaking, I can conceive of this being the case, and if all the particles of matter attract each other by virtue of a hidden force which resides in them, I can comprehend without difficulty the vast amount of motion diffused throughout the universe; but from the very moment that I cease to speak of matter as inert, I am obliged, on the other hand, to say that it is active, since I acknowledge that it encloses a principle of action. Here we find ourselves face to face with an immense difficulty; and it will be said, doubtless, that every scientific man since Newton has been obliged to solve it, since it is impossible to leave at the very foundation of Science two contradictory assertions. In fact, minds habituated to scientific pursuits know that it is necessary to seek outside of bodies the cause by which they tend towards each other; they know that in enunciating the law of universal gravitation, they regard results, and not causes, and only mean that things take place as if bodies were attracted in direct ratio of their mass, and inverse ratio of their distance. Such is the reservation which all sensible minds have made, or ought to have made, more or less explicitly.

Now, what light is this new theory going to throw upon the principle of gravity? Here is the answer. A substance to which the name of ether has been given is diffused throughout the entire universe. It envelops bodies, and penetrates into their interstices. The existence of this substance is deduced from a series of proofs, among which luminous phenomena hold the first rank. Ether is composed of atoms which impinge upon each other and upon neighboring bodies. It forms, in this way, a universal medium, which exerts a constant pressure upon the molecules of ordinary matter. The new theory accounts for the reactions which are produced between the ethereal atoms and the material molecules; it proves that these reactions are such that the material molecules must tend towards each other precisely according to the conditions which the law of gravity also observes. We shall attempt farther on to give an idea of this ingenious demonstration. For the present we leave aside all the proofs, and only declare results. Every one will understand the importance of the point to which we have just arrived. It becomes evident that bodies do not owe their gravity to an intrinsic force, but to the pressure of the medium in which they are immersed.

The motion of heavy bodies would no longer appear to us other than as a transformation of the ethereal motions, and gravity, henceforth, enters into that majestic unity to which we have conducted all physical forces.

Thus heat, light, electricity, magnetism, cohesion, chemical affinity, gravity, are all resolved into the idea of motion. All these motions may be converted into each other according to fixed relations, some of which are known, but by far the greatest number of which is, as yet, undetermined.

Let us see if the idea of matter will not henceforth be rendered more simple and clear. At the foundation of our system we have the atom of ether. But is there, — we supposed it just now for the sake of making ourselves more easily understood, — is there really an ether, and an ordinary matter, differing from ether in its essence? To speak more clearly, are *there two kinds of matter?* We can hardly conceive it, now that we have resolved everything into motion. In what respect would these two kinds of matter differ? Would not the one be subject to the same laws of motion as the other. Can there be two systems of mechanics? Certainly not; since there is but one law for motion, there can be but a single essence for matter, and the molecules of ordinary matter must *appear to us as aggregates of ethereal atoms.* It is under this form that we shall represent the elementary particles of simple bodies, of iron, lead, oxygen, carbon. The molecules of these bodies do not differ in their substance, but simply in the interior arrangement of the ethereal atoms which compose them.

Because iron, lead, oxygen, carbon, chemically unite in different combinations, must we suspect in them some difference in substance? Upon what would this difference depend, since chemical affinity itself presents to us only the idea of motion?

II.

The Hypothesis of the Unity of Natural Phenomena in its Relation to Science.

FROM the point now reached, we may consider, in all its bearings, the hypothesis whose principal features we have just traced. If it be admitted in its entire force, natural phenomena present themselves under a form so simple as to surprise the mind at once with wonder and awe. The physical world is composed of atoms of one kind. By virtue of a motion received and communicated to each other by these atoms, they become so grouped and intermingled as to form simple molecules, compound molecules, gaseous, liquid, and solid bodies. It is to one and the same cause,—namely, to motions received and converted into others,—that we must attribute molecular aggregations in the realm of the infinitely small, and in that of the infinitely large the gravitation of the heavenly bodies. It is this motion of a fixed nature which, either in bodies or outside of them, constitutes the phenomenon we call heat; it is the same motion which, under another peculiar form, constitutes light, under another electricity, and so on.

The atom and motion! Behold the universe!

Upon this basis will the mathematician be able to construct his calculations. While applying his equations to a medium composed of uniform atoms, and seeking all the motions which may be produced, and all the combinations that may spring from these motions, he will come again to the recognized phenomena of physical science, the laws of the planetary circulation, of the propagation of sound, of luminous undulations. Entered upon this path, he will determine by means of the analogies which such a study will suggest, in addition to motions known and recognized, motions which appear probable. He will here find again, doubtless, the laws of matter studied already; he will here find, perhaps, properties towards which the attention of man has never been directed. How many important laws thus reign around us without our even suspecting it! How long men lived without a suspicion of the electric phenomena whose action encircles them! What unexpected revelations may spring up from this study of nature from a new point of view!

Here let us speak only of the obscurities which the new hypothesis has already dispelled, and let us leave to the future the task of justifying the hopes to which it has given rise. By means of the bond which it establishes between all natural phenomena, our minds are accustomed to seek, in every fact, through the transformations which formerly obscured our vision, its immediate origin and its direct result.

When we see a steam engine raise a weight, or overcome a resistance, we think at once of the coal burning in the furnace, whose combustion effects the work of the machinery. But where does the coal get this power which we know how to utilize? It is the product of a long-continued work of the sun, stored up in fossil vegetables. Thus all facts are brought, as it were, to a common standard, and we become accustomed to look always for a due proportion between cause and effect.

To give a familiar form to thought, may we here

cite an anecdote? We will borrow it from Father Secchi, who relates it in his work upon the unity of physical forces. There was, in 1855, at the Universal Exhibition in Paris, a huge bell, of enormous weight; it was held up by a system of props so ingenious that a single man was able to keep it in motion; only the tongue had been removed, without doubt out of regard for the ears of visitors. The man who exhibited the bell easily kept it swinging, and the spectators admired the facility with which he made this formidable engine move. An ecclesiastic, an educated and intelligent man, — we may suspect that it was Father Secchi himself, — approached the exhibitor, and said to him, —

"Your system of supports is very well contrived; it permits you to set this huge mass in motion with extreme facility; but would it be the same if the bell had its clapper, and if it struck?"

Those standing by doubtless did not understand the thought of the jocose ecclesiastic.

The fact really was, if the bell had been forced to give out sounds, — that is to say, to make the air vibrate strongly, — he would have failed with his best efforts to find the necessary strength to produce such a vibration. However perfect had been the mechanism of the support, this strength would have been borrowed from the arm which pulled the rope. When a bell vibrates, it is the labor of the bell-ringer which is turned into sound. To remove the tongue—that is to say, to prevent the sound—is to render the task easy to the ringer.

As is the cause, so is the effect. We find ourselves placed at this point of view whenever we attempt to recall, briefly, the main facts upon which rests the unity of physical forces. Before entering upon the examination of this, we desire to reply to two questions which arise about the hypothesis we are developing.

Is this hypothesis a useful one?

Is this hypothesis really new?

Firstly, is it useful? The great advancements made by modern science are due to experiment and observation. *Non fingo hypotheses*, wrote Newton as preface to his works. *Nullius in verba*, proclaimed the escutcheon of the Royal Society of London. *Provando e riprovando*, was likewise the inscription upon the shield of the Florentine Academy, founded by Galileo. Certain it is that modern physical science has done away with examining phenomena themselves, independently of their supposed causes, by subjecting them always to exact instrumental measurements. We can to-day appreciate with the utmost exactness the thousandth part of a millimetre (a millimetre = .039 inch), the ten thousandth part of a second, and we are not likely to give over the researches which such means of investigation enable us to undertake. Surely the experimental method, which has given birth to such brilliant results, is not likely to perish, and it is always of the vernier of the physicist, the scales of the chemist, the scalpel of the physician, the telescope of the astronomer, that we must ask for sure information in regard to Nature.

Is this to say, nevertheless, that, while this incessant work of research is pushed in all directions, we may not attempt to group together facts already discovered, in such a manner as to rise to more and more general laws? Vain would be the attempt to combat this tendency of the human mind. It is easy to say that one must concern himself only with proven facts, and leave the rest to dreamers; but it is not so easy to keep to this programme. Every one is irresistibly led to form for himself an idea of the universe as a whole, be it correct or not. Among men who make real progress in science, those even who appear most absorbed in searching after particular facts, those who confine themselves to the patient investigation of particular phenomena, certainly have their general theories, which they forbear, perhaps, to communicate to the public, but which guide them in their labors, which induce them to attack one question rather than another, which, true or false, suggest new ideas to them, and classify their difficulties.

Far above all the theories which have thus guided men of science, now rises this grand conception of the unity of the physical forces. It is only an hypothesis, but it offers itself with guaranties sufficiently sure to necessitate a kind of revision of science as a whole. It will illuminate with new light facts already known, trace out a new path for researches in questions hitherto beset with difficulty and doubts, and point out in what direction it is necessary first to question Nature. Were the hypothesis false, experience would know how to profit by it.

But, it will be said, is it not to be feared that, led away by this seductive theory, many observers will come to pay but little attention to facts, and strive to force them into the outline they have sketched in advance, and thus unwittingly falsify the results of their experiments before presenting them to the public? This will doubtless happen — it has already happened; but it is not a very serious evil, for science is well enough armed against such a danger, and erroneous assertions cannot long resist its control.

But, it will be again objected, scientific men are not

the only ones concerned in the matter. Your hypothesis lays hold upon philosophy. Not only does it comprise all physical science, but it also enters the domain of metaphysics. Philosophers will doubtless adopt it, believing that they hold in their grasp a scientific truth, and they will find, perhaps, that they embrace only a chimera! What reply do you make to that? The metaphysicians must have a care to keep themselves well informed.

Now, is this veritably a new hypothesis, which presents the physical world to us as composed of uniform atoms and diverse motions?

Properly speaking, there are but few ideas which can be brought forward as entirely new. If we confine ourselves to mere definitions and the surface of things, we shall discover the theory of the unity of the physical forces in the remotest antiquity. The philosophers of ancient Greece did not have a single fact, scientifically demonstrated, at their disposal, so to speak, and in this state of things they formed the most simple hypothesis concerning Nature. They had a clear field, and nothing interfered with their empiricism; so they went straight to the most general conceptions, and every one in his own way made unity out of the grand total. Thales of Miletus, six hundred years before our era, began by declaring that water was the principle of all things. Fifty years later his countryman, Anaximene, beheld in air "the uniform and primitive element." The Eleatic school, in Magna Grecia, sought elsewhere the universal principle. "Nothing proceeds from nothing, and nothing can change," said Xenophanes; "everything possesses the same nature;" nevertheless he demanded, in order to account for the multiplicity of variable substances, two elements, water and the earth. About the year 500, Heraclitus adopted fire as the unique principle and universal agent. "The world is neither the work of gods nor of men; it is an ever-living fire, enkindling itself and extinguishing itself according to a certain order." Here, then, are four elements successively proclaimed — water, air, earth, fire; and by a kind of eclecticism, they came to be admitted, all four at once, into the composition of the universe. Aristotle accepted these four elements, and for long ages after him they served as the basis for every system of nature. During the eighteenth century the four elements were still admitted, on the eve of the great works which have founded modern chemistry.

Pursuing this general progress of ideas, we encounter the Atomic theory itself in very ancient times. Leucippus, an Eleatic, who lived five hundred years before our era, conceived the universe as formed of a vacuum, and a real substance, the last division of which was the atom. "The round atoms," he said, "have the property of motion. It is the combining and separating of these which give birth to things and destroy them. All physical phenomena are determined by the order and position of the atoms, and only take place by virtue of necessity." Democritus of Abdera, a disciple of Leucippus, developed his doctrine. He attributed to atoms, similar to each other, original properties, impenetrability, and a sort of weight. For him "every active influence or every passive impulse is a motion following contact."

He distinguished impulsive (παλμός) and the movement of reaction (ἀντιτυπία), whence results the circular or whirling motion (δίνη). In this consists the law of necessity (ἀνάγκη) pointed out by Leucippus. Epicurus, the Athenian, adopted the views of Democritus, and made a kind of atomic theory. He gave to the atoms a hooked form, and supposed them endowed with an oblique motion in relation to each other, in order that they might be able to grasp each other, and form bodies. Such is the system of which Lucretius sang in his magnificent poem of Nature.

But, need we repeat it? the conceptions of these philosophers, of these poets, were purely Utopian. Formed outside of facts, they brought no light into the domain of physical science; their authors could see in them only what they themselves had put there, that is, the caprice of their imagination; thus they did not possess for them the meaning which they now have for us. In their eyes they were but simple formulæ, which they little cared to confront with the facts of Nature, and which served only as preambles to their systems of philosophy.

What we say of the ancients wholly applies to the middle ages, to the period of the Rénaissance, to the primitive works of modern times. The physical system of Descartes has scarcely more value than that of Epicurus; the same fantasy, the same vortices, the same hooked atoms.

The great men who, at the time even of Descartes, inaugurated the remodelling of the sciences, concerned themselves only with facts, and left aside hypotheses. Such was Kepler, such was Galileo. When the second generation of great savans came, the generation of Newton, of Leibnitz, of Huyghens, so rich was the store of precise information that a general hypothesis was well nigh impossible. Science was divided into several branches. In each one of them they made one or more particular hypotheses; but it was a long time before they thought to include in one general formula the numerous and exact phenomena which a diligent and laborious search was bringing to light.

If, now, from the mere examination of these phenomena a general formula arises, if a system springs up spontaneously from the study of observed facts, we may pronounce it veritably new, even if its formula should be ancient, if even there might be found in Democritus an almost complete enunciation of it. The originality of the hypothesis brought forward at this time consists in this, that it has the support of a considerable number of facts; that it has its birth in these facts. It borrows its worth from the facts it embraces; it becomes, in a certain manner, a fact itself.

III.

The Difficulty encountered by endeavoring to express new Ideas by Means of Old Terms.

THE theory we are investigating will only appear in its true light when we have examined some of the phenomena upon which it rests, and pointed out the new aspect which it gives to certain portions of science.

We have no intention, as may be thought, of giving a course of lectures on Physics. We shall only be able to touch upon certain points, to throw out a few hints.

Do not ask of us a general picture of Nature, when we only seek to sketch a few details of it; through these partial openings may doubtless be seen what would be the scope of the work which we do not dream of undertaking.

Furthermore, we shall adopt, in our cursory survey of natural phenomena, the same order we have pursued in our summary exposition of the system. We shall speak first of what relates to light, heat, electricity; we shall next come to that other group of actions, chemical affinity, cohesion, gravity, the principle of which current prejudices more especially locate in the very bosom of the molecules.

Heat! Electricity! Cohesion! Gravity! we say. These words even bring us to make a declaration, the advantage of which will accrue to us during the whole course of this essay.

In every branch of physical science, we just now said, particular hypotheses have been made. They have influenced the language which has been adopted in the various branches of science. In many cases the names given to the phenomena, the classification of them even, are in disagreement with the new theory.

What are we to do under these circumstances?

Undoubtedly for a new situation there is needed a new language. But shall we create here this new language complete in all parts? We have many other difficulties to encounter.

Shall we have recourse to paraphrases in order to avoid words which seem to contradict the idea we are unfolding? We should run a great risk of not being understood.

We shall continue, then, to call all things by their customary names; if, however, this nomenclature be found in disagreement with our fundamental idea, we beg that the accident may be kindly attributed to the state of transition through which physical science is now passing. In former times the electricians admitted the existence of a positive and a negative fluid. They recognized, henceforth, in a current a positive pole and a negative pole; we shall do as they have done, without its leading to material consequences. When a body is heated without being permitted to expand, it absorbs, in order to acquire a certain degree of temperature, a determined quantity of heat, and if it is heated and also allowed to expand, it requires, to arrive to the same temperature, a greater quantity of heat. Physicists have given to the excess of heat demanded in the second case the name of latent heat of dilatation. We may continue to call it latent, while seeing clearly that it is used up in producing the mechanical work of dilatation.

In studying the molecular actions, there have always been recognized attractive forces and repulsive forces; we may do the same, prejudging nothing as to the existence of these forces.

As to the word Force itself, we preserve it for want of a better in our vocabulary. Every time a motion appears to us as the continuation or the transmutation of another motion, we can do without the idea of force, and we ought to reserve this notion for motions, the origin of which remains entirely concealed from us. We shall continue, nevertheless, as we have done in the preceding pages, to employ the word force in its customary sense. We shall speak, without scruple, of the force of gravity which causes a stone to fall, and of the force of cohesion which maintains a body in the solid state, at the same time supposing that the fall of the stone and the solidity of the body are due only to movements of the surrounding medium.

To speak truly, the inconvenience we here point out is not a new one, and these difficulties of language are well known in physics. As in each of the parts of this science different hypotheses have been successively made for the purpose of grouping and co-ordinating the phenomena, so physicists have learned, in a certain measure, to withdraw themselves from the empire of words, to deal abstractly with the ideas which a common signification awakens; they know how to see facts beneath the conventional picture of them which words give.

Nevertheless, the explanation we have just entered upon is not useless; it will justify the want of harmony that will often be found, without doubt, between the names given to phenomena and our mode of appreciating them.

CHAPTER II.

SOUND AND LIGHT.

I.

Nature and Mechanical Equivalent of Sound.

It has been known for a long time that sound is the effect of the vibration of bodies, propagated either through the air or some other medium. Acoustic phenomena are, so to speak, visible to the naked eye; their nature was likewise known at an early period. If a plate of copper be firmly secured by one of its sides, and a bow drawn across the free edge, the eye perceives the vibration of the plate. Again, if the parchment of a drum, upon which fine sand has been scattered, be exposed to agitated air, the disturbance of the sand betrays that of the air, and its grains, driven from the parts that are most shaken, are seen to collect along the lines when the air and parchment are in a state of rest. The rapidity of the propagation of sound is itself easily appreciable by the senses. Everybody knows that if a cannon be fired off at a distance, the light is seen much sooner than the report is heard; so that it can be easily ascertained how many seconds sound requires to pass through a given interval of space. Open to direct experiment, the principles of acoustics have long been viewed in their true light, and it has not been necessary to imagine, for their explanation and orderly arrangement, either a special fluid or a particular force. Sound is seen to be a vibratory movement, produced by a certain impulse, and propagated through a medium. There has not been introduced into physics either a sonorous *fluid* or a sonorous *force*.

We can say, then, but a few words concerning sound. Let us observe, however, that the study of sonorous vibrations, considered in its relations to the history of science, possesses a peculiar interest for us. These are the first vibratory movements which were well understood, and on the day in which they were exactly defined was laid one of the firmest foundations of the new physics. The facts revealed by this study aided in a powerful manner those great minds which established the theory of light. Between sonorous and luminous vibrations many analogies have been sought for and found. Great dissimilarities have also been encountered. Here is one of the most important, which we mention here, although it is soon to be referred to again. The sonorous vibration takes place in the direction of the propagation of the sound; each molecule of the air that has been set in motion executes a to-and-fro movement along the same line in which the sound is propagated. On the other hand, the luminous vibration takes place in a direction perpendicular to the ray of light. The points of resemblance and dissimilarity revealed by the study of the motions of light and sound afford us at the outset a primary view of the problems which the new physics encounter, and of the methods it is able to employ for solving them.

We have still another instance of the kind of investigation demanded by this new science when we propound a question apropos of sound that will successively arise in regard to all physical phenomena. We have said that these various phenomena are susceptible of being transformed into each other, and we are thus led to look for a common measure in the dynamic effect which they represent. What is the dynamic effect of a sound? or, to employ a term introduced into the language of science by the study of heat, what is the mechanical equivalent of sound? Let us take a bell, and strike it with a hammer; we shall be able to calculate exactly the mechanical work due to the stroke of the hammer. This will be a certain number of kilogrammetres.* The bell will vibrate, and we shall be able to measure the amplitude of its vibrations by means of a luminous ray reflected upon a small mirror attached to the bell; this is a method often employed in acoustics to amplify the oscillations, and to render them visible. If we make in this manner a series of experiments, and if we compare the figures which express the shocks with those which express the amplitude of the oscillations, we shall be able to condense the result of this examination into a formula which will give us an idea of the sonorous effect of different blows.

But shall we in this manner obtain a true mechanical equivalent of sound? Shall we be able to say that the unit of sound is equivalent to so many kilogrammetres? To do this, it would be necessary to begin by determining the value of a unit of sound. We distinguish in sound several properties; it has

* The kilogrammetre is the work represented by a kilogramme raised to the height of a meter. The English equivalent of this term is "foot-pound." The quantity of heat necessary to raise one pound of water one degree Fahrenheit in temperature, is competent to raise a weight of seven hundred and seventy-two pounds a foot high. By the French method of reckoning, the quantity of heat necessary to raise a kilogramme (2.2 pounds) of water one degree Centigrade in temperature, is sufficient to raise a weight of four hundred and twenty-five kilogrammes to the height of a meter (39.37 inches). — *Translator.*

pitch, which depends upon the number of vibrations, intensity, which depends upon their amplitude, quality, which depends upon more complex conditions. What phenomenon shall be selected for comparing different sounds with each other, while keeping account of all their effects? Hitherto such a question does not seem to have demanded consideration. Interest has been felt alone in the number of the vibrations upon which musical theories depend.

To tell the truth, we do not see that there is practically any special utility in selecting a sound unit which shall correspond to the conditions we have just indicated. We shall not insist, then, upon this point, and we have only mentioned it in order to show one of the new aspects presented by physical studies.

It has always been one of the chief difficulties of the exact sciences to determine suitably the units by which phenomena must be compared. This selection of units possesses now an especial importance from the new standpoint at which physicists are placed. We are thus confronted here with a capital question, which demands some elucidation. If we have only touched upon it with regard to sound, it is because we await an opportunity to treat it with more profit, for we must necessarily encounter it at nearly every step of the way.

II.

The Nature of Light and of Interference.—Universality of this last Phenomenon.

Acoustics teaches us that sound is a vibratory motion, be it of air, water, or of any other material medium analogous to these. In examining optical phenomena, we shall presently behold ether appearing as the agent of the luminous wave, and this conception of ether will soon become, as it were, the bond of all the ideas which pertain to the unity of physical forces.

Let us take a prism composed of two plates of glass, separated by sulphide of carbon, and place it in the path of a beam of solar rays, and receive the image of this beam upon a screen. This image, as is known, is called a spectrum; the screen will show us luminous rays of different colors, unequally refracted by their passage through the prismatic mass of the sulphide of carbon. The red rays are least deflected, and are consequently found towards the edge of the prism; then, proceeding from the edge to the base, come orange, yellow, green, blue, indigo, violet.

If, now, we examine the spectrum with attention, the phenomenon will not be simply a luminous one.

It will acquaint us with the calorific and chemical properties of the solar beam. Let us receive the spectrum upon a plate pierced with a narrow slit, through which the rays can act upon a thermo-electric pile, and let us move the slit through the whole extent of the spectrum, beginning with the violet portion. So long as we remain in the violet, the indigo, the blue, and even the green, the needle of the thermoscopic apparatus will be deflected but slightly. It will indicate a heat, increasing according as the slit crosses the yellow, next the orange, then the red; but let us pass beyond the red, and enter the dark part of the spectrum; we shall here find the maximum of heat.*

Thus there is, beyond the visible image of the solar beam, a warm spectrum which we cannot see. If the rays, which are refracted on one side of the spectrum beyond the red, have an especial aptitude for producing heat, those which are refracted upon the other side, beyond the violet, have an especial aptitude for exciting chemical action. These chemical rays may be rendered visible by a contrivance well known in the

* From an examination of the distribution of heat in the spectrum of the electric light, it appears that the maximum of intensity is in the *dark* region, beyond the red; and further, that if all the visible rays were conveyed to a focus, its heat would be only one ninth of that produced at the dark focus of the invisible rays. — *Translator.*

laboratory. Take a sheet of paper, the lower part of which is moistened with a solution of sulphate of quinine, while the upper part remains dry. Let the image of the solar ray fall upon this sheet, the spectrum preserves at the top of the sheet its ordinary appearance, while in the moistened portion a brilliant phosphorescence appears beyond the violet rays.

Thus the spectrum extends beyond its visible portion in two directions, to the right and to the left, and analysis can distinguish in it, beyond the luminous rays, calorific and chemical rays, the latter more particularly deviated towards the violet portion, the former more especially refracted towards the red portion.

All forms of light thus far known exhibit three kinds of rays. Their phenomena vary, it is well known, in a certain degree with the means of observation. And in the first place, simply to use a prism, produces a spectrum in a manner merely conventional; the prism disperses differently the rays of different refrangibility; it leaves the red rays more crowded together; on the contrary, it gives more breadth to the violet portion. We may, by other means, obtain a spectrum in which the different rays better preserve their relative value. The nature of the prism also changes the relation between the luminous, calorific, and chemical rays. If a solar ray be received upon a prism of water, the maximum of heat will appear in the yellow, upon a prism of common glass in the red, upon a prism of flint glass beyond the red, upon one of rock salt far beyond the red in the entirely dark portion. It would likewise be necessary to take into account the nature of the luminous source. But passing over these details, we were desirous only of showing how, in every emission of light, there is found, in addition to luminous action, properly so called, calorific and chemical action. We succeed in dividing these three actions, but not without difficulty, to such a degree do they appear to be commingled. Let us not, then, forget this ready-made synthesis, which is offered to us at the outset. If, after having studied separately heat, light, and affinity, we arrive at the law which unites these phenomena, let us remember that we found them united, and that we have separated them in order the better to examine them. For the present we must continue our analysis, leaving aside heat as well as chemical action, and occupying ourselves only with light.

What is light? This subject has given full play to the imagination of the earlier physicists. Some located in the eye a visual force; this force projected rays which came in contact with objects. Others supposed, on the other hand, that objects emitted all around them an infinite number of little images, which entered the eyes of men and of animals. It was hardly possible to discuss seriously the nature of light before the structure of the eye was known, and the image of objects had been seen formed upon the retina, as upon the end of a *camera obscura*. The retina thus impressed transmits the sensation to the optic nerve. But how is the retina thus impressed? how is the image formed there?

Newton supposed that luminous bodies shoot out little corpuscles, the shock of which excites the retina. This is the famous theory of emission, which gave rise, towards the close of the seventeenth century, to such hot disputes. Newton had established, while making use of his hypothesis, the principal laws of optics, those of reflection and those of refraction. Nevertheless difficulties continued to exist. Other optical phenomena, more complicated, — polarization and double refraction, — could not be explained by the Newtonian theory. Questions were put to Newton, to which his hypothesis gave no answer: "Where does the light go when it is extinguished? Whither go the corpuscles which are constantly leaving the sources of light?"

Descartes advanced the idea that a subtile matter fills the planetary spaces. This conjecture, by which he had vainly attempted to explain astronomical phenomena, was eagerly seized upon, and applied to light. Malebranche was among the first to suspect that light is produced by the undulations of an ether, and that the differences in the length of the waves are the causes of the different colors. Huyghens adopted this system, and subjected the deductions to mathematical calculation. Thus admitted into science under a hypothetical title, the existence of the ether became more and more probable in proportion as experiment justified the conclusions drawn from this principle.

Nevertheless, Newton supported with energy the theory of emission, and accumulated in its defence proofs, a great many of which would appear very whimsical to-day. Euler supported Huyghens, and beheld, in a kind of classification of the phenomena which affect our senses, an argument in favor of undulation. "In order to perceive an object by touch," said he, "it is necessary that we be in contact with the object itself. With regard to odors, we know that they are produced by material particles which issue from the volatile body. As regards hearing, there is nothing detached from the sounding body. The distance at which our senses recognize the presence of objects is nothing in the case of touch, small in the case of smelling, somewhat great in the case of hearing, while in the case of sight this distance becomes considerable. Pursuing this progression, we must believe that the sense of sight perceives in the same manner as that of hearing, and not in the same manner as the sense of smell; it must be supposed that luminous bodies vibrate like sonorous bodies, instead of emitting particles like volatile substances."

During the debate there were brought forward some curious facts, observed near the middle of the seventeenth century, by Father Grimaldi, a Bolognese monk, who left behind him a very original treatise upon optics (*De Lumine, Coloribus, et Iride*: Bologne, 1665). If a shutter be pierced with a very small hole, and the luminous cone which passes through the orifice be examined, it is observed that the cone is much less acute than would be expected, considering only the rectilinear transmission of the rays. The experiment becomes still more striking if there be interposed in the path of the luminous ray a second shutter pierced with a new hole, when it is readily apparent that the rays of the second cone are more divergent than those of the first. If a fine thread is introduced into the luminous cone, and its shadow projected upon a screen, the shadow appears surrounded by three colored fringes, and there are also seen in this shadow one or more luminous rays. Let the image of the orifice in the shutter be received upon a screen, and a white circle is seen surrounded by a dark ring, next a white ring, more brilliant than the central portion, then a second dark ring, and finally another very faint white ring. If in the shutter with which the experiment is made, two very small holes are pierced at a distance from each of one or two millimetres, and the two images received upon a screen in such a manner that they overlap each other, it is found that in the lenticular segment formed by the overlapping of the images, the circles are more obscure than in the part where they are separated. Thus it appears that by adding light to light darkness is produced.

These curious facts, minutely described by Father Grimaldi, appear to us quite decisive, now that we grasp their real meaning. It seems to us that they should have caused the system of undulations to have triumphed forthwith; but even those who could appreciate their value in the seventeenth century, were far from drawing from them all their consequences. These experiments at least served to feed the dispute. Corpuscles, said Huyghens, coming directly from the sun, and passing through a small aperture, would form, in escaping from the holes, a straight cylinder, and not a cone. The conical form is proof of a motion which is propagated in a line lateral to the luminous ray. Newton retorted, "if light is a motion, it would not remain confined in a narrow cone; it should spread itself in every direction, and scatter itself in a circular manner around each point of disturbance."

"Without doubt," replied Huyghens, "at every point of the luminous ray spherical undulations go out in the direction lateral to this ray, and extend into all the surrounding space; but they are not often enough repeated to produce the sensation of light. They do not yield to a force as strong as do those which occur in the same direction as the ray, and they destroy each other in their confusion." *

The first scientist who saw all that could be inferred from these experiments of Grimaldi was Thomas Young, that sagacious traveller, who developed several branches of physical science, and who discovered the key to the Egyptian hieroglyphics. The researches of Young were continued by Arago and Fresnel, and more recently by M. M. Fizeau and Foucault. The labors of all of these have given a complete explanation of the *fringes* of light pointed out

* In accounting for the fact that light is not diffused beyond the rectilinear space when it passes through an aperture, Huyghens says, "Although the *partial* waves produced by the particles comprised in the aperture do diffuse themselves beyond the rectilinear space, these waves do not *concur* anywhere except in front of the aperture." — *Translator.*

by Grimaldi, and the theory of *interferences* which they have founded is one of the most glorious achievements of modern thought.

The principle of interferences is difficult to grasp. A ray of light, according to what we have just said, is the propagation of a motion in which the atoms of the ether oscillate around their point of equilibrium. They are then endowed with a certain velocity in one direction during the first half of this undulation, and with the same velocity in the opposite direction during the second half. Let us suppose, now, that we can arrange two rays issuing from the same surface, and that by any contrivance whatever one of the two has been retarded behind the other a half undulation, if two rays be placed upon each other at the point of superposition, the atoms of either will remain immovable, since they will be equally solicited to motion in both directions; there will then be at this point an absence of luminous motion, or darkness. There will be an increase of light when the amount of retardation is two demi-undulations, darkness when it is *three*, and so on.

By means of experiments based upon this principle, it has been possible to measure the length and the duration of the waves which correspond to the different colors of the spectrum. The wave increases in length and in duration from the red to the violet; its length, expressed in millimetres, is 0.000738, at the extreme red; 0.000553, at the middle of the yellow; 0.000369 at the extreme violet.*

It has been proved, moreover, by particular methods, that the same law of decrease extends to the invisible portions of the spectrum; the calorific vibrations beyond the red are slower and longer. The longest wave of obscure heat which it has been possible to measure up to this time is 0.001830 millimetres. As regards the duration of the waves, a general idea may be obtained from knowing that the vibration of the yellow ray lasts 530 trillionths of a second. It is, moreover, a recognized fact that the eye cannot perceive a sensation of light unless it continue at least several hundredths of a second; there must be then several billions of waves to produce the sensation of light.

We here see confirmed by experiment the reasoning which we just now put in the mouth of Huyghens,

* According to Tyndall, the length of a wave of mean red light is about the 39,000th of an inch; that of mean violet light, the 57,500th of an inch. Taking the velocity of light at 185,000 miles per second, as determined by Foucault, we have the number of waves of red light which enter the eye each second as 458,142,400,000,000. The number of waves which enter the eye to cause the sensation of violet color is about one third more than this, being upwards of 700 trillions. — *Translator.*

and the limits within which the waves once departed from the line of luminous shock, are no longer frequent enough to produce light.

It may be understood, without dwelling further upon this point, how important the study of interferences becomes in the new physics. The interest pertaining to it does not remain restricted within the limits of optics, it extends to every branch of science. Whenever a vibratory motion exists we may expect to find the phenomena of interferences.

Acoustics, for example, has its own, which easily admit of proof. Let us take a plate of copper, supported horizontally upon an upright stand, and having scattered fine sand over it, let us draw a bow rapidly over one of its edges. The surface is divided into eight triangles; the adjacent triangles vibrate in an opposite direction, while those which do not touch each other vibrate in a similar direction. The fine sand gives evidence of this state of things by arranging itself along the lines which cross the surface. There is here a state of repose, because there is an equal tendency to two opposing motions. These contrary impulses, which go out from the different parts of the metal plate in order to cross each other in the surrounding air, must produce therein true phenomena of interferences, for sometimes they mutually strengthen each other, and sometimes they oppose each other. This is demonstrated by a very simple instrument — a tube, one end of which forms a funnel over which a membrane is stretched, while the other extremity terminates in two branches, forming an angle with each other. Having now the ear placed against the funnel, the two terminal branches of the tube are passed over the surface of the plate of copper, when it is easily observed that the sound is very weak when they are near the centre of two contiguous triangles, and that it grows louder, on the contrary, when they touch two triangles vibrating in the same direction.

We are now acquainted with sonorous interferences and luminous interferences, but above all we must expect to see these phenomena of universal occurrence in physics. They will appear necessarily under the most varied forms, according to the mode of motion which produces them, and according to the nature of the organ whose office it is to perceive them. In every case, the researches which will be made in this direction will be powerfully aided by the magnificent works that have marked the study of luminous interferences.

III.

Conclusions as to the Ether drawn from Luminous Phenomena.

We must now consider more closely this idea of the Ether, to which we have been led by the phenomena of light; we must clearly define it, and remove it from the controversies to which it has given rise.

What is the ether? Is it really imponderable? and in that case what does this property mean? In what does it differ from ordinary matter? in what does it resemble it? What are its relations with it? Is it not a little strange to introduce here, at the very time when we are banishing from science a host of conventional entities and abstract forces, the idea of a medium, which is, so to speak, immaterial?

We shall have replied to this last question when we shall have shown that the ether, according to our conception of it, does not possess the fantastic properties that are sometimes ascribed to it.

We conceive of a simple gas, oxygen for example, as an assemblage of elementary molecules, animated with motion, which strike against each other, from which result the expansive force of the gas, and the pressure that it exercises upon the bodies in which it is contained. This idea will become clearer when we seek to inform ourselves concerning the interior constitution of bodies, profiting by the ideas generally adopted regarding the nature of heat; but for the present we can accept it as a kind of primitive conception, with which the mind may be satisfied while waiting for the testimony of science. It is under this simple form that we conceive of ether, and we add, that its elements are atoms; that is to say, that they cannot be divided. If the objection be offered that it is difficult to comprehend that they are really indivisible, we reply, that it is sufficient for us to conceive that they comport themselves as such, for no one has the pretension to penetrate either the infinitely small or the infinitely great. The atoms of the ether are endowed with motion, which they communicate to each other, and to surrounding bodies.

Are these, then, immaterial? Certainly not. Two properties belong to matter — impenetrability and inertia. The ethereal atoms are impenetrable at the outset; they are so by definition. They are also inert; they have received the motion with which they are endowed, and they lose it only in communicating it. Nothing distinguishes ether, then, from matter; and when we just now presented it, in the preceding lines, as awakening the idea of a medium, so to speak, immaterial, we made, be it understood, a pure concession to certain usages of language. Our ether is material, just as oxygen is.

But it is imponderable! Yes, and we here confront a very delicate explanation. We should make ourselves better understood if we should here show, in some detail, under what aspect universal attraction appears in the new order of ideas upon which we are entered; but this is a point of view which we shall unfold only in the course of this work. Under whatever form the interior state of an ordinary molecule may be conceived, whether it be regarded as a primitive substance, or be viewed as a reunion of ethereal atoms associated together according to certain laws, it must be admitted that this molecule possesses a mass much larger than each of the atoms of the ether. This granted, if two molecules are in the presence of each other, the surrounding ether impinging upon them both in every direction, there will result from this very situation a disposition to approach, which is known by the name of attraction or gravity. Let us remain satisfied for the present with this summary explanation, which will be completed in the sequel. It is sufficient now to make it evident how ether is imponderable; if two molecules tend to approach each other, it is because their presence interrupts the uniformity of the ethereal impulsions in just such a manner that they are necessarily forced one against the other. Nothing of a similar character will be discovered when we examine ether itself in its own motion ; it moves in all directions, and there would appear to be nothing to force it in one direction rather than in another. Thus this fluid produces attraction in matter without itself being subjected to it ; it confers gravity upon bodies, and itself is imponderable.

If, then, we wish to distinguish ether from ponderable matter, it will be necessary, in order to employ an accurate term, to call it imponderable matter. Granted that the current phrase be ether on the one hand and matter on the other, we shall still continue to use the term, as we have done already, for the sake of brevity ; we shall have at least shown what is expressed by these words, and we shall have proved that the imponderability of ether must be admitted, without meaning thereby to confer upon this fluid a title of immateriality. Let us add that there would be a real advantage in getting rid of the term ether, which incurs the risk of always possessing more or less of mysticism.

We have represented the ether as an assemblage of atoms which strike each other, and rebound in all directions. Here a capital objection presents itself, and we must attack it. How do these atoms rebound? Are they elastic? The notion of an atom and of elasticity are incompatible. We can understand the elasticity of a compound molecule; the different parts of the molecule, pressed upon by an external force, are displaced while being compressed, then regain their position while returning the impulsion which they have received. This mechanism supposes a void in the interior of the molecule; but the atom is impenetrable, indivisible; it does not enclose a void. There is here a serious difficulty. Huyghens, it is necessary to state, ascribed to the atoms of the ether an elastic force. What did he understand by this? Did he then regard them as compound corpuscles? The difficulty had only changed place. Fortunately mechanics have succeeded in solving this problem, and the beautiful researches of Poinsot upon revolving bodies explain how the ethereal atoms may rebound from each other without being elastic.* It is sufficient for understanding this effect to suppose that

* "It is just as easy to conceive of a vibrating atom as to conceive of a vibrating cannon-ball; and there is no more difficulty in conceiving of this Ether, as it is called, which fills space, than in imagining all space to be filled with jelly. You must imagine the atoms vibrating, and their vibrations you must figure as communicated to the ether in which they swing, being propagated through it in waves; these waves enter the pupil, cross

they possess, in addition to their motion of translation, a rotatory motion.

From the theorem formulated by Poinsot, it results that a hard and inelastic body may, if it revolves, be turned aside by any obstacle, precisely like a body possessed of elasticity; more than this, there is often, after the blow, a greater velocity than before, because a part of the rotation is converted into a motion of translation. In general, when a revolving body strikes against an obstacle, it cannot lose its two motions at once; at farthest, it may do so only in certain theoretical instances, which need not be considered here. If the shock pass through the centre of gravity of the body, it will assist its onward progress, but not its rotation; if it be eccentric, it will stop the rotation, but not the forward movement. The two motions will likewise be partly transformed, one into the other, in such a manner as to produce more varied phenomena.

The game of billiards has made some of these effects familiar. It is here seen how the rotation of a ball operates to modify both its direction and velocity when a blow is received. In the illustration cited, elasticity combines with rotatory motion; but it is

the ball of the eye, and break upon the retina at the back of the eye. The act is as real and as truly mechanical as the breaking of the sea-waves upon the shore."—*Heat as a Mode of Motion*, p. 268.

necessary to abstract the latter phenomenon, and to give it separate consideration, in order to conceive how the ethereal atoms may rebound without being elastic.

Let us enter a little farther into the idea of these motions ; we shall see the hypothesis of the rotation of ethereal atoms explain, in a certain degree at least, a phenomenon of capital importance, and one we have already mentioned.

The undulation of light, we have said, is propagated in a direction at right angles to the luminous ray, and we have observed that in this respect it differs from the sonorous undulation which takes place in the same direction as the propagation of the sound. There is nothing to astonish us in the manner in which the luminous wave is propagated, and we find many examples of it in nature. If a stone be thrown into the water, we see the water rise in waves which are perpendicular to the direction of its fall. In this case it is evident that the liquid disturbed by the stone moves in the direction in which it meets with the least resistance.* A similar reason was assigned by Fresnel

* "In the case of sound, the vibrations of the air-particles are executed *in* the direction in which the sound travels. They are therefore called *longitudinal* vibrations. In the case of light, on the contrary, the vibrations are *transversal*, that is to say,

for the luminous motion. "I think," said he, "that the shock is communicated to the ether longitudinally, that is to say, in the direction of the ray, but the ether possesses such a nature that it can only respond to the impulse by a lateral vibration." This vague explanation becomes strikingly definite if we suppose the ethereal atoms to turn upon themselves.

We know from mechanics, that if a revolving body receive a shock perpendicularly to the axis of rotation, the centre of gravity of the body is carried at right angles to the direction of the shock. Strike a whirling top, it will jump to one side. There is a well-known experiment upon this subject. A top is placed on a horizontal plane, and while it *sleeps*, if the plane be inclined from south to north, immediately the top moves from east to west; if the plane be inclined from east to west, it moves from south to north. Thus the component gravity causes the top to move in a direction at right angles to that component. The phenomenon does not take place, of course, unless the top is whirling, and there is no such result while it is in a state of repose.

the individual particles of ether move to and fro *across* the direction in which the light is propagated. In this respect waves of light resemble ordinary water-waves more than waves of sound." — *Tyndall, Fragments of Science*, p. 284.

Placed at this point of view, we understand without difficulty how the rotation of the ethereal atoms accounts for their lateral displacement during the luminous impulse; their transverse vibration will appear not only possible, but even necessary.

This explanation we borrow from the books of Father Secchi upon the Unity of the Physical Forces. The inference which the learned abbé has drawn from the rotation of the ethereal atoms, is not the least interesting feature of his work. While we are dwelling with some detail upon the transverse motion of light, we cannot resist the desire of presenting, in opposition to the hints thrown out by Father Secchi, the views which M. de Boucheporn offers upon the same subject in his *Principe général de la Philosophie naturelle.* This digression will interrupt the orderly presentation of our ideas; but we shall at least acquaint our readers, by a brilliant example, with those bold conjectures characteristic of M. de Boucheporn, which he knew how, with infinite art, to verify in fact. Seeing his point of departure and the good to which he arrives, one is persuaded, but not convinced.

M. de Boucheporn attributes the transverse undulation to the friction of the ether against the revolving surface of the sun. This hypothesis furnishes him at once with the explanation of the phenomena of colors, and he finds its confirmation in the examination of the length of the waves which characterize the principal tints of the spectrum. Let us follow him in his argument.

If it is the sun's rotation which impels the ethereal atoms in a direction tangent to its motion, this effect should be produced in a very different manner at the different points of the solar meridian ; it would necessarily decrease in energy from the equator to the poles. At the equator the friction is in full force, while at the poles it is nothing. Between the equator and the pole its energy decreases according to the radii of the parallels, or as the *cosines* of the latitudes. M. de Boucheporn, thereupon, supposes the differences in the lengths of the waves, that is to say, the differences in the colors, correspond to the impulses given at the different parallels. What will be the parallels which characterize the different colors ? M. de Boucheporn immediately searches for those which offer remarkable peculiarities, those whose trigonometrical lines, the *sines* and *cosines*, have the most definite relations with unity. He finds eight, and he assigns to each of them one of the tints of the spectrum, the red being placed at the equator. In this manner he draws up the following table, in which the *cosines* of the latitudes selected will be found opposite to the tints which are attributed to them :—

		Cosines of the Solar Latitudes.
Violet,		0.33
Indigo,		0.50
Blue,		0.60
Green,		0.70
Yellow,		0.80
Orange yellow,	Fresnel had taken these two instead of the orange alone.	0.87
Orange red,		0.93
Red,		1.00

It is next to be ascertained whether these numerical values are proportional to the length of the waves, the determination of which has been made by Fresnel with such admirable precision. M. de Boucheporn observes here in his hypothesis, that the experimental values of Fresnel represent the sum of two effects; the progressive motion of the sun exercises a friction as well as its rotation. The first of these two effects may be eliminated by cancelling a constant number in the values given by Fresnel, and then these values take the following form, the length of the red wave being taken for unity:—

	Lengths of Undulation.
Violet,	0.396
Indigo,	0.518
Blue,	0.600

	Lengths of Undulation.
Green,	0.696
Yellow,	0.800
Orange yellow,	0.865
Orange red,	0.932
Red,	1.000

If the two numerical series which precede be compared with each other, there will be found between them the most perfect harmony that could be asked for experimental calculations.

The views of M. de Boucheporn present a striking aspect, if we consider the three principal colors of the solar spectrum, the blue, the yellow, and the red, which are alone able to make white light without the aid of the intermediate tints. For these three fundamental colors, the values expressed in the two series are exactly as the numbers 3, 4, and 5; and not only do these numbers present a very simple relation to each other, but they are also the only ones which satisfy, in a simple manner, another characteristic condition: the square of one of them is equal to the sum of the squares of the other two: $9+16=25$. This is what M. de Boucheporn calls the *law of the three squares.* It takes a prominent part in his theories, and we cannot help appreciating its importance. The motions which strike our senses are the better grouped the more simple are the numbers which express them ; at the same time the intensity of our sensations is in a direct ratio with the squares of these numbers. Our senses, then, are called upon to judge of the double condition which is fulfilled when these numbers and their squares present such very simple relations. We can here see, with M. de Boucheporn, one of the harmonies of Nature.

The general explanation that has just been given upon the subject of the transversal motion of light will doubtless be regarded as but a brilliant flight of fancy, and our chief object in mentioning it has been to justify the judgment we passed at the beginning of this essay upon M. de Boucheporn's book. On the other hand, the hypothesis of the rotation of atoms which we have borrowed from Father Secchi, is a very plausible and suggestive one, and it would be necessary to be on one's guard against putting the two conceptions upon the same footing.

Whatever may be the reason assigned for the transversal motion of the luminous wave, the fact itself is certain. It has been fully proved by the phenomena of polarization.

When a ray of a single color, a red ray, for example, is reflected by a plate of glass at an angle of thirty-six degrees, it acquires from this single circumstance peculiar properties.* If to this reflected ray a second mirror of glass be presented at the same angle of thirty-six degrees, and the mirror is made to turn in every direction around the point of incidence, it is observed that the ray is no longer reflected with the same intensity in all directions. There is a plane in which the reflection is greatest, and one in which it is almost nothing. The maximum takes place in the plane parallel to the plane of reflection upon the first mirror, and is consequently called the plane of polarization; the minimum occurs in the plane which makes a right angle with this. If, instead of selecting a ray of a particular color, just as now mentioned, we

* When a luminous beam impinges at the proper angle on a plane glass surface, it is polarized by reflection. It is polarized, in part, by all oblique reflections, but at one particular angle the reflected light is perfectly polarized. An exceedingly beautiful and simple law, discovered by Sir David Brewster, enables us readily to find the *polarizing angle* of any substance whose refractive index is known. This law was discovered experimentally by Brewster, but the Wave Theory of light renders a complete reason for the law. A geometrical image of it is thus given: When a beam of light impinges obliquely upon a plate of glass, it is in part reflected and in part refracted. At one particular incidence the reflected and the refracted portions of the beam are at right angles to each other. The angle of incidence is then the polarizing angle. It varies with the refractive index of the substance, being for water 52½°, for glass 57½°, and for diamond 68°. — *Fragments of Science*, p. 257.

employ white light, we obtain similar results, only a little less distinct, because the angle of incidence under which they are produced varies a little for the different colors.

What is, then, this modification which the ray undergoes when placed in the conditions we have mentioned? Why does it no longer behave like an ordinary ray? The transverse vibration furnishes us with the reason. Before the luminous beam falls upon the first reflector, the waves are propagated around it transversally in all directions. They diverge around this axis as the spokes of a wheel go out from the hub. At the moment of its falling upon the mirror, the glass absorbs one portion of the waves and reflects the other. What are chiefly the ones it sends back? Those which are parallel to its surface, and therefore which can less easily penetrate it. If we go to extremes in order to render the phenomena more intelligible, we shall consider the reflected ray as no longer containing other than parallel waves between them and the surface of the first mirror. The ray, then, is said to be polarized, and this term, though invented by Newton for an altogether different hypothesis, explains sufficiently well the fact. What will now happen when these waves, conducted in a single direction, shall fall upon a second surface of glass?

They will be wholly reflected at the moment that the mirror is parallel to them, and they will, on the other hand, be absorbed more and more in proportion as this mirror is made to revolve. Such is, in its essence, the phenomenon of polarization, and it appears that it may be explained without difficulty, by taking for the starting-point the transversal undulation.

Fresnel has even shown that if two rays, polarized at right angles, happen to coincide, they give no sign of interference, even when there is between them the difference of a half undulation. This will be understood by recurring to the fundamental notion of interferences, and it will not be a matter of astonishment that vibrations, when they are perpendicular to each other, do not destroy each other, as happens in other cases.

If we could push this investigation a little farther, we should be able to show how the data assumed in the matter of luminous motion have, one after another, received striking confirmation. The principles being laid down, mathematical analysis has developed their consequences, and observation has justified these results. In pursuing this twofold task Fresnel made for himself a glorious name. His calculations, his experiments, are alike memorable; one hardly knows what most to admire, the high intelligence with which he has presented the facts, or the practical skill with which he has verified them. In no other portion of science has man so nearly arrived at the secrets of nature, or submitted its fundamental phenomena to such exact measurements.

IV.

What does the Study of Light teach us concerning the Molecular Constitution of Bodies?

In pointing out some of the laws of optics, we have endeavored to give form and substance to the notion of ether. It has been seen how the motions of this fluid have been analyzed and measured. It has been seen that the ether, notwithstanding its imponderability, possesses the properties of matter. Henceforward, if we resume the thread which is to guide us, if we place ourselves at the stand-point from which all physical phenomena appear as exchanges of motions, we are led to ask if it has been possible to precisely define the conditions under which the atoms of the ether exchange their motions with the ponderable molecules. Diffused throughout the stellar spaces, enclosing among its particles the celestial globes, the ether thus penetrates into the deepest recesses of all bodies, and bathes their ultimate molecules. Thus there is no phenomenon in which it does not play a part, either the principal, or, at least, a secondary one. If, then, we could know the mass and velocity of the ethereal atoms, and the mass and the velocity of the ponderable molecules, we should possess, in some sort, the key to the physical sciences. He, at least, who should discover some bond of union between these; who should be able to grasp, in some degree, their relation; such a one would open up a fruitful source of discoveries.

Is there need to say it? No such discovery has been made up to the present time. We establish by the results the reciprocal action of ether and of ordinary matter, we see an incandescent body produce light, we see this light converted into chemical action; but in no instance do we know how to reduce the phenomenon to its mechanical elements, and to seize the moment itself of the exchange of motion.

In regard to the distances themselves of the atoms from each other and of the molecules from each other, we have only rough and contradictory estimates. It is generally supposed that the spaces left between the ponderable molecules are enormous in proportion to their dimensions. Thomas Young did not hesitate to affirm that the molecules of water are placed in relation to each other as if a hundred men were equally distributed over the whole area of England, that is to say, distant from each other thirty English miles. Nevertheless, crystallographers are far from believing in such considerable spaces. As far as ether is concerned, Cauchy has deduced, from very delicate calculations, that the distance of the atoms approximates to the two hundredth part of the red wave: according to this, there would be three hundred thousand atoms in the length of a millimetre. M. de Boucheporn, on his side, believes he can affirm that the ethereal atoms are crowded together to such a degree that the sum of the empty spaces is reduced to the twentieth part of the sum of those filled by the atoms. In conclusion, these problems remain intact, and their solution has not even been approached.

It would seem that the light which passes through bodies ought to give us information in regard to the molecular spaces.

There are transparent substances, the molecules of which permit the luminous waves to pass freely, without losing any part of their motion. Among transparent bodies, a certain number are colored; they arrest or absorb the waves of certain colors only. Thus a solution of sulphate of copper allows the blue waves to pass, and stops the red rays; if there be projected upon the screen through this solution a spectrum, the red end of the spectrum is entirely cut off. A piece of red glass, on the other hand, owes its color to the fact that its substance may be freely traversed by the red waves, while the shorter waves become extinguished in it; if it be placed in the path of a luminous beam, the spectrum is reduced to a band of lively red. If there be placed, at the same time, in the path of the beam the solution of sulphate of copper and the piece of red glass, these two transparent bodies extinguish all the rays at once, and produce a complete opacity. Some other body, a solution of permanganate of potash, for example, will extinguish at the same time the red and the blue rays, and give passage only to the yellow, which constitute the central portion of the spectrum.

Different bodies, then, exercise in relation to luminous waves a sort of power of election, extinguishing some, permitting others to pass. Here it is the longest, there the shortest, which are arrested; in other cases both the longest and the shortest are stopped at the same time, while those only of medium length can obtain a passage. Whence comes this difference? What law presides over this sort of select choosing on the part of the luminous rays? No doubt it arises from the form of the molecules, and the nature of their motions. We are hardly able to say more about it. The difficult molecular motions seem to have rhythmical periods peculiar to each, in virtue of which they come into harmony with those of the ethereal atoms.

That there should thus be in molecular motions a kind of rhythm, from which results the selection of colors, has been demonstrated in a general manner by the study of the spectrum ; but there may be mentioned in this connection several curious and characteristic facts, which have been pointed out in the last few years.

Project upon a screen the spectrum of a solid body strongly heated. So long as the body remains incandescent only, so long as its molecules are not freed from the bonds of cohesion, the spectrum remains continuous ; there are seen neither dark nor bright lines ; the waves of all the colors and of all the intermediate shades are produced at the same time. If the body be heated still more, it passes the point of incandescence, and enters a state of combustion ; then the molecules become free, at least for a moment. Then, also, the bright and dark lines appear in the spectrum. The waves are consequently re-enforced in some places, and weakened in others ; they are governed by a new law.

That it is the molecules themselves of the heated body which, in their state of freedom, impress upon the waves these peculiar modifications, cannot be doubted, for every substance gives in this manner lines so clear and definite that their appearance alone is sufficient to distinguish it.

Acoustics furnishes us, upon the subject of these phenomena, with analogies which present themselves spontaneously to the mind; that which takes place when the body is incandescent may be compared to the noises which result from waves mixed up and of every length; the effects produced by the free molecules remind us of the harmonious sounds emitted by strings, the vibration of which is not impeded by any obstacle.

Here is also a new fact recently discovered.

We have just seen that every substance in burning gives its own lines. When, for example, sodium is burned, a very bright line is seen in the yellow portion of the spectrum, in a clearly marked locality. (Line D, of Fraunhofer.) If, now, instead of burning sodium, we interpose the vapor of sodium in the path of the ray which should give a continuous spectrum, the phenomenon is completely reversed; at the exact point where just now there was a bright line, a dark line appears. Thus the vapor of sodium, when it acts as a screen, absorbs exactly those rays which it emits when it acts as the luminous source.

This fact observed in the case of the vapors of iodine, of strontium, and of iron, has become generalized; it is now known under the name of the *reversement of the spectrum;* it shows that bodies tend at the same time to absorb and to emit the same waves.

Shall we be astonished at this double tendency, viewed from our present stand-point, and shall we not recognize in it the necessary consequence of the principles which explain, in our opinion, all physical science? From the moment that certain ethereal motions have a special facility of becoming converted into certain molecular motions, these latter must also easily undergo the opposite conversion. The reciprocity of the motive forces insures to us that of the phenomena.

If the natural bond of union of all the facts we have mentioned be sought for, it will be seen that, taken together, those facts come to the support of that grand law we have endeavored to expound, and which we have designated under the name of the Unity of the Physical Forces; but there will at the same time be observed the defects in the method we are obliged to employ. Were we dealing with a system ready-made, we might unfold it step by step, and pass from one part to another without a gap. Far from this, we have to do with a system seen only in glimpses, scarcely sketched, even, the elements of which are so incomplete as to be found insufficient, it may be. Such being the case, what remains for us to do, if not to show some parts brightly illuminated, while leaving in the shade whatever is obscure? From those scattered lights, these fugitive glimmerings, must result the conception of the whole.

The digression we have just made with reference to acoustics and optics has exhibited to us the branches of science in which the phenomena of motion have been best studied—that motion which we now see underlying all things. Sonorous motions, luminous motions, have been verified, measured, and scrutinized in all their modes; but, on the other hand, their mechanical effects have scarcely been glanced at. The study of heat will afford a contrary result; heat-motions are at the most suspected only, and continue to be but little known as regards their real nature; but their mechanical effects have been demonstrated by splendid experiments, and measured with the utmost precision.

Sound and light on one side, heat on the other: here are two subjects of study, as yet incompletely explored ; but these two studies complement each other, and as soon as we approach them we see shining brightly forth that idea by virtue of which Nature appears to us but as a system of mutually exchanging motions.

It is from comparisons of this sort that our thesis derives its principal strength. This is what our readers must never lose sight of while we continue to confront the system of the unity of the physical forces with facts which experience has acquainted us with.

CHAPTER III.

HEAT.

I.

The Dynamic Theory and Mechanical Equivalent of Heat.

THE theory which reduces the physical world to matter and motion, presents an external so attractive as to provoke a sort of mistrust. Accustomed as we are to complicated appearances, we are astonished at this pretentious unity. We ask, with uneasiness, if we are not the dupes of a desire to simplify everything. Is not this hypothesis, which gives us, in a manner, an insight into the whole plan of Nature, a deceitful mirage? Are we not deluded by fallacious generalizations? Are we not enticed into falsifying the phenomena for the sake of forcing them to enter into the frame of a preconceived theory? To these questions a reply can be made only by an examination of the facts, and for this purpose we have taken a brief survey of the different provinces of physics.

The natural sequence of this work brings us now to investigate heat, and we thus find ourselves led to discoveries which have served as the origin of the theory of the unity of the physical forces.

Here our task becomes easy, perhaps, but, it must also be added, somewhat thankless. The equivalence of heat and mechanical work has been for some years past not a new idea; books, public instruction, and lectures, have spread it abroad; the popular mind has been zealously informed of it. We have not to fear, then, that the subject will be a strange one to our readers, or fear, rather, lest they have heard too much of it, and that they regard it as commonplace. We shall then only recall, very briefly, the principles of thermo-dynamics, and we shall strive more especially to throw light upon the consequences that may be drawn from them as regards the constitution of bodies.

Let us mention, in the first place, a book which presents in a distinct and agreeable form all the essential data of the new theory of heat. It contains twelve lectures, delivered by Mr. John Tyndall, at the Royal Institution, London, upon *Heat, considered as a Mode of Motion.* The course was delivered in 1862; the book appeared, in France, translated by the Abbé Moigne, in 1864, and it was immediately appreciated at its true value by all persons who are interested in the general advancement of science. It is impossible to give to lessons in physics a greater charm, and, at the same time clearness, than has been done by Mr. Tyndall in the work we name. The book has preserved the form of oral instruction; we follow the words and gestures of the professor; we assist in the details, even in the very mishaps of the experiments. It must not be thought, however, that we have before us an impromptu effort reproduced by the art of the stenographer. Much skill lies concealed under this apparently easy method of procedure. Mr. Tyndall skilfully calculates all his results; the accidents in his experiments occur only when well foreseen; they are fortunate mishaps that take place only when he wishes to seize the attention of his public, whether listeners or readers, and abruptly direct their thoughts to some striking anomaly.

The experiments of Mr. Tyndall are, moreover, conceived with much skill and ability. For a long time he has been a master in the art of lecturing before a numerous audience. He has contrived ingenious instruments to magnify the results of experiments. He was one of the first to employ the electric light for projecting on screens the enlarged images of the most delicate phenomena. The dramatic effects which have made the success of Mr. Tyndall's lectures, appear adroitly preserved in his book.

As to the substance of these lectures, the professor treats his subject bit by bit; he takes his time to produce in the minds of his pupils those ideas which the regenerated study of heat awakens.

"Remember," said he to them, "we are entering a jungle, and must not expect to find our way clear. We are striking into the branches in a random fashion at first; but we shall thus become acquainted with the general character of our work, and, with due persistence, shall, I trust, cut through all entanglement at last." When he has sketched the principle of a new conception, "Do not be disheartened," he makes haste to say, "if this reasoning should not appear quite clear to you. We are now in comparative darkness, but as we proceed light will gradually appear, and irradiate retrospectively our present gloom." And, in another place, "Whenever a difficult expedition is undertaken in the Alps, the experienced mountaineer commences the day at a slow space, so that when the real hour of trial arrives, he may find himself hardened instead of exhausted by his previous work. We, to-day, are about to enter on a difficult ascent, and I propose that we commence it in the same spirit; not with the flush of enthusiasm, which the necessity of labor extinguishes, but with patient and determined hearts, which will not recoil should a difficulty arise." The professor conforms in every particular to his excellent programme, and employs a good deal of skill in preparing his pupils for the abstract notions he wishes to impart to them.

Nevertheless we cannot refrain from adding that the conclusions of the book remain vague and unsatisfactory.

We are acquainted with the history of the works, and of the discoveries which successively modified and defined the idea of heat.

As in the case of light, two theories for a long time confronted each other: that which made of heat a material substance, and that which saw in it only a mode of motion. The material nature of caloric continued to be admitted much later than that of light. During the latter years of the eighteenth century, Lavoisier and Laplace, in presenting to the Academy of Sciences a memoir compiled in common, upon the subject of heat, seemed to hold the balance equally between the two opinions. Their language was, "We shall not decide between the two foregoing hypotheses; several phenomena would seem to favor the latter (that of motion), as, for example, the heat produced by the rubbing together of two solid bodies;

but there are others which are more simply explained by the first (that of its material nature); perhaps both obtain at once."

In reality they abandoned the idea of motion without having made any use of it, and returned to the theory of the material nature of heat. Laplace, especially, after the termination of his connection with Lavoisier, became a confirmed supporter of this last theory, which was thus strengthened by a weighty authority.

A little later, during the last years of this century, Rumford, an original, almost paradoxical mind, resolutely pronounced himself against the materiality of heat. "If heat," he said, "is a matter lodged in the pores of different substances, it may be forced out, as water is squeezed from a sponge, and the same body will not be able to emit it indefinitely." Having thus brought the question to a decisive experiment, he caused an iron bar to turn upon another similar bar in the midst of a liquid, and he showed that there was a disengagement of heat as long as the iron bar was turning.

The experiments of Rumford had not as much celebrity as they deserved. Thomas Young alone seemed to understand their bearing. In a treatise upon physics, published in 1807, he exhibited the labors of Rumford, and compared them with his own discoveries in regard to heat; but the old ideas upon caloric continued to hold their sway in men's minds.

Steam engines came, and all the questions pertaining to heat again became the order of the day. At this time the materiality of heat was so little disputed that Sadi Carnot took it as the basis of his celebrated Reflections on the Motive Power of Fire (1824). It is known how, starting from this erroneous doctrine, Sadi Carnot, and his celebrated commentator, Clapeyron, revived the thermo-dynamic theory. They called attention to the causes which enabled an engine burning charcoal in its furnace to produce work at its shaft. They had this good fortune, that their arguments—their formulas, even—could be disengaged from the fundamental error which contaminated them, and serve as a basis for the new theory of heat.

In 1839 M. Seguin published an Essay on the Influence of Railroads. In it heat was considered as a motion, and the author gave very appropriate hints concerning the transformation of this motion into work; but this subject was but touched upon in the book of M. Seguin, who had particularly in view questions of social economy.

Between the years 1840 and 1850 were produced the remarkable works of M. M. Mayer and Joule.

Starting from very different premises, and placed at totally different stand-points, — the one in Germany, the other in England, — they came at the same time to clearly demonstrate the equivalent of heat and mechanical work, and they determined the ratio of this equivalence.

Immense result! It was like a shining beacon lighted up in the midst of the darkness which enshrouded physical science when this definite fact was made known. A unit of heat is equivalent to four hundred and twenty-five kilogrammetres, or, in other words, the quantity of heat which is required for elevating one degree the temperature of a kilogramme of water, can also do the work which consists in elevating four hundred and twenty-five kilogrammes to the height of one metre.

This discovery has for fifteen years vastly increased the field of vision for science. There springs from it, as it were, a new philosophy of nature. A mental revolution is taking place, of which we see only the commencement, and we are endeavoring to sketch the beginnings of this change.

All uncertainty in regard to the nature itself of heat came to an end as soon as its mechanical equivalent had been determined. What is it that could be transformed into motion in so regular a manner if it be not another motion ?

Doubtless there would not be discovered at once, either in the action of steam engines, or any other phenomena, the precise mode of the transformation ; but the principle of it was grasped by the mind with conviction. The motion itself was not seen, but its effects were both perceived and measured.

Heat is a motion ; but of what kind ?

Some physicists conceived at first that it might be due to the longitudinal vibrations of the ether. They knew that ether, by its transversal vibrations, produced light. With regard to the longitudinal action, that which is produced in the direction of the ethereal ray, no especial property was known, and they disposed of it by attributing to it the calorific effects. This conjecture, which did not rest upon any well-grounded fact, has gathered around it but a very small number of supporters, and has scarcely been considered more than a work of the imagination.

According to present notions, heat is a motion of the molecules themselves of bodies. All material molecules are endued with this motion ; they strike, without cessation, against each other, and thus maintain or change their state. By means of their shocks the molecules of bodies cause us to experience the sensation of heat, and from the intensity of these shocks we determine degrees of temperature. This perpetual vibration of the molecules itself constitutes the phenomenon of heat; but it may naturally convert itself into different effects. It can, when circumstances are favorable, agitate the ether, and produce light; it can agitate the air, and produce sounds; it may concentrate itself in order to move masses, and produce what has very properly been called mechanical work.

Properly speaking, the different effects we have just mentioned — heat, light, sonorous shock, mechanical work, and other effects of the same class, which we do not mention at this time, — are but the different manifestations of the same cause. The motion with which each molecule is endowed at a given moment constitutes for it a sort of intrinsic energy. In mechanics we are able to appreciate and measure the energy with which a moving body is endowed. The product of the mass of a body into the square of its velocity expresses what is called the *vis viva*, or living force. This product has not, properly speaking, any physical representation, and it offers to the mind at first a conception rather abstract; but it assumes a capital importance from this circumstance, that it is equivalent to double the work that the body can produce in losing all its velocity ; it gives then the measure of the dynamic effect which the body in motion contains within it. We may now declare, making use of this idea, that all particles of matter possess, at a given instant, a certain amount of living force. They may lose a portion of it in doing some work, — that is, by displacing a mass, — but then the *vis viva* which they lose becomes stored up in the work performed, and it is renewed when this work is undone.

Let us consider a steam engine, and neglecting all the losses of force or work which belong to the mechanism itself, let us think only of the theoretical or ideal action of the engine. The steam expands, forcing out the piston ; each molecule of vapor thus loses a certain quantity of *vis viva.* These accumulated losses produce a revolution of the shaft, which is engaged, for instance, in elevating a weight. At the end of the operation all the *vis viva* which the steam has lost is virtually found in the elevated weight. If I cut the cord which sustains this weight, it will fall, and reproduce in its fall all the *vis viva* which has been expended in order to raise it. It will appear in the form of heat at the instant when the body strikes the ground, and if this could be collected and restored to the steam, the latter would be replaced in the condition in which it was found at the beginning of the operation.

What we have indicated in this rough example, is constantly taking place in all nature. To bring the *vis viva*, or living force, into the condition of work, and then to reproduce it, in this lies the whole activity of Nature.

II.

Changes of State produced by Heat furnish Information as to the Constitution of Bodies.

ADMITTING an incessant agitation of the molecules, it is easier to account for the phenomena which take place in bodies when they pass from the gaseous to the solid and liquid states.

As a general rule, all bodies are susceptible of these three states : carbonic acid gas has been liquefied and solidified ; water appears to us under the form of ice and vapor ; we know how to fuse and volatilize the metals. We do not always possess sufficient means for making every body pass successively through these three states ; but we may, nevertheless, affirm that we should see them under all three forms, if our researches could include a sufficiently extended scale of temperatures.

As a general rule, also, it may be said that heat must be increasingly added to the same body to bring it from the solid to the liquid state, then to the gaseous. Thus heat triumphs over the bonds which bind together the molecules; it combats those attractive forces which manifest themselves in the interior of bodies, and which have preserved until now so mysterious an aspect.

Has it been possible, through the antagonism which is displayed between heat and the attractive forces, to isolate the calorific motion? to separate it from the phenomena which mask it? to determine its special mode and laws? Unfortunately not. Nevertheless it may be said that the study of gases has thrown much light upon this question.

How must we conceive of the gaseous state? To begin with, it is characterized by a considerable distance between the molecules. Endued with a great projectile velocity, these molecules hurl themselves against each other, or against the bounds of the space which confines them. Have they merely a projectile motion? They have necessarily a rotatory motion also, for, if this motion did not exist at a given moment, it could not fail to be generated by the incessant collisions of the different molecules. The eccentric shocks, those which do not pass through the centres of gravity, are, in fact, of such a nature as to produce a rotation. This rotation combines with their elasticity to cause the molecules to recoil from each other. The former alone might produce this effect, if the molecules, instead of being compound, were but simple atoms. A sort of medium state is thus established in the gas. If the motion became weak at certain points, it would at once be strengthened by the agitation of the rest of the mass. Moreover, each molecule recoils without definite direction, and may go in all directions, to be successively projected into every part of the entire mass. There is a state of complete liberty.

Let us observe that the molecular distances are considerable; their velocities are also considerable. What becomes, then, of that effect which must be produced at the instant when two molecules approach and strike each other, that effect which is attributed to the attractive forces, whatever they may be? This effect is, so to speak, annulled; it lasts only a very short time relatively, since the molecular distances are very great. It is but a very transient effect, since the velocities are enormous. It becomes so weak that it may be neglected; thus, in the gases, the attractive forces possess no power. The calorific motion exists in them without antagonism, and may be observed in its integrity.

If we apply cold to a gas, if we cause it to lose a portion of its *vis viva,* the energy and amplitude of its oscillations will gradually diminish. A moment will come when each molecule will be, as it were, imprisoned by its neighbors, and forced to oscillate along a limited curve. The gas will have become a liquid.

From the very fact alone of the proximity of the molecules the attractive forces have regained the supremacy, and have destroyed, in part, the mobility of the system. Gravity, too weak before, now makes itself felt, and the molecules are obliged to arrange themselves in such a manner as to present a surface parallel to the horizon.

Along this surface they are detained in their new position by one of their sides only; upon the other side their motions remain free, and they possess an especial aptitude for returning to their former state, and an evaporation takes place from the surface.

Moreover, in the remainder of the mass, the molecules still enjoy a relative amount of liberty. They are enclosed within restricted orbits, but their axes of rotation continue to lie in all directions. They can thus roll over each other to some extent.

Besides, the bonds which limit their movement yield to the slightest effort, and the whole mass may be mixed without difficulty.

Let us continue the application of cold. The molecules approach each other; they enter, one may say, within the sphere of each other's action, and they remain there; their axes of rotation become upright, and take a common direction; the body has passed into the solid state.

During these conditions the molecules still oscillate; but they can no longer, without outside assistance, depart from the circle in which they are kept by their neighbors.

In describing the manner in which bodies change their state, we have just brought forward the attractive forces. After the repeated declarations we have already made, we might almost forbear to mention that these forces are for us but the symbols under which lie concealed the ordinary phenomena of motion.

Before terminating this work, we shall be brought to consider collectively these attractive forces, which we admit this time by way of inventory. Nevertheless, let us now make a rapid allusion to them, so as not to leave them entirely in mystery.

It is a necessary consequence of the rotation of molecules that they draw with them into their motion a certain number of ethereal atoms. They are thus wrapped in a sort of atmosphere whose radius may vary according to circumstances, and which nearly represent that which we just now called the molecular sphere of action. So long as these atmospheres do not touch each other there is no action; such is the case with gases. If the molecules approach each other, and the atmospheres slide over each other (this is the case with liquids), action begins, an action purely mechanical, due to the meeting of the ethereal atoms. If, finally, the atmospheres penetrate more deeply into each other, the effect is more strongly pronounced; the ethereal envelopes that are penetrated find themselves hindered in their progress, and they behave so as to render parallel the rotations of the different molecules respectively, as happens in the case of solids.

With this sketch, let us pass hastily on, as we are unwilling to delay, in order to speak of liquids and solids, the constitution of which continues even now very obscure. It is sufficient that we have shown how the laws of this constitution are connected with those laws, far better known, which govern the gaseous state. Thanks to this solidarity, the gases offer us a convenient type for studying molecular motion, and we may fix our attention upon them some moments longer, sure of deriving from them information applicable to all the forms of matter.

III.

Theory of Gases.

THE theory of gases, the principle of which was just now pointed out, has been much studied of late years, and it has given rise to a great number of remarkable publications. It does not present itself, however, as an entirely novel conception, for its fundamental notion could be found in the Hydrodynamics of Bernouille, published in 1738; but, buried in the work of Bernouille, it scarcely saw the light until the last thirty years, and it has received its development only through the quite recent works of M. Joule and M. Clausius.

We cannot here follow those two physicists in the analytical deductions, by means of which they have defined, with wonderful accuracy, the theory of gases. But we shall at least be able to show how the hypothesis in regard to the constitution of gases, which we have just sketched, accounts for the facts successively revealed by experiment.

From the simple enunciation of this theory arise, as a necessary consequence, several of those celebrated laws which form the very foundation of physics.

It results primarily from our hypothesis, that the

molecules of a gas may be considered as every instant moving in a straight line, with a uniform velocity common to the whole mass; we have, in fact, eliminated the perturbing phenomena existing at the moment of collision. Is it not henceforth evident that, if the gas is confined in a receiver, the pressure exerted upon its walls will be proportional to the number of atoms in the gas, that is to say, to its density? Equality of ratio between pressure and density, such, we know, is Mariotte's law.

Now, at the same pressure and temperature, different gases of the same volume contain the same number of molecules. This is a fact which chemists particularly bear witness to, and it is deducible from our hypothesis. Since the molecular actions, properly so called, may be neglected, it is conceivable that the molecules of the same gases, endowed with equal liberty, will arrange themselves, other circumstances being equal, at equal distances from each other, and in the same volume of a gas the same number will be found. A quart of hydrogen, a quart of oxygen, a quart of nitrogen, thus contain a uniform number of molecules. What will take place if two gases are mixed? The same principle applies to the mixture, no special action resulting from the proximity of the molecules, since the nature of the molecule appears to have no influence upon the phenomenon. Atmospheric air will behave in this respect, like pure oxygen or pure nitrogen. This is the law of gaseous mixtures pointed out by Gay-Lussac.

Since the distance between the molecules remains the same, whatever their mass, it is to be expected that the same quantity of heat will be necessary in all the gases, to raise the temperature of the elementary molecule one degree. It will be objected, perhaps, that the heaviest molecules will receive from this quantity of heat a less velocity; this is evident; but it is also evident that they require a less velocity for manifesting that effect which we call an elevation of one degree of temperature. We are then brought to this result, that the temperature of the elementary molecules of the different gases is elevated one degree by a like quantity of heat, whatever may be their mass, or, as chemists say, their atomic weight. Under this form may be recognized a celebrated law, to which Dulong and Petit have given their name.

Gay-Lussac, it is known, has established that the coefficient of expansion is uniform for all gases. Now, have we not here a natural result of the facts we have just disclosed? These molecules, which are all placed at the same distance from each other, and which absorb the same quantity of heat in order to increase their temperature one degree, ought they not to separate from each other equally under this increase of temperature? The experiments of Gay-Lussac have shown that the coefficient of this uniform expansion is $\frac{1}{273}$ of the primitive volume.*

This examination might be continued; but we have said enough to show how, from our very definition of gases, flow the laws which characterize the gaseous state.

The laws of Mariotte, Gay-Lussac, Dulong, and Petit have had a singular fate. Discovered at a period when methods of experimenting were far from the perfection to which they have since attained, they were at first regarded as absolutely exact, and applicable, in full force, to the different gases. When that advance of improvement in the methods of experimenting, to which in France is attached the name of M. Victor Regnault, took place, these laws, till now so respected, were at fault in numerous cases; they fell into suspicion; at least they came to be considered as empirical formulæ, which represented in only an approximate manner the general course of phenomena. No theoretical conception, in fact, accounted for the

* Decimally expressed, this *coefficient of expansion* becomes .00367, according to the Centigrade scale.

numerous perturbations which the exact methods employed by scientific men gave evidence of. But now we see why gases obey but imperfectly Mariotte's law, and those other laws we have just referred to. In order to establish them, we have been obliged to suppose that every molecule might be considered as constantly endowed with a uniform and rectilinear motion, and we have regarded as insensible the duration of the periods in which this motion was interfered with. If this duration becomes appreciable, while at the same time remaining very slight, the arguments which we gave cannot be repeated in their full force. Here may be seen the source of much of the distrust in the old laws: it may even be seen how the perfect gaseous state is, to a certain extent, but an ideal which is scarcely realized in fact. Hydrogen would seem to attain to it entirely; oxygen and nitrogen, and, as a consequence, atmospheric air, come near to it; but carbonic acid is sensibly removed from it. As to vapors, they behave like gases only as they are very far from the point of liquefaction.

There are, then, but very few perfect gases; but they furnish us with valuable information, in showing us matter entirely disengaged from those attractive forces which complicate molecular phenomena.

When we heat a cubic metre of air under a constant

pressure, all the living force which the gas receives is employed in increasing its volume by $\frac{1}{273}$ for every degree of temperature.* When, instead of leaving the pressure constant, we prevent the gas from expanding, when, while heating it, we force it to remain enclosed in the space of a cubic metre, all the living force acquired by the air is employed in increasing its pressure $\frac{1}{273}$ for every degree. If the initial temperature is that of melting ice,† at two hundred and seventy-three degrees, the pressure of the air is doubled.

The same law holds true below zero. If, instead of heating the gas, we cool it, its pressure goes on diminishing $\frac{1}{273}$ for every degree. If we could attain to two hundred and seventy degrees, the gas would no longer possess any pressure; it would be but an inert mass of molecules, deprived of all living force. This is what has been called the *absolute zero* of temperature. There is here a sort of limitation to which it is not possible to attain in practice, and at which all molecular motion would cease.

We have been considering a definite mass of air,

* That is, $\frac{1}{273}$ for one degree *Centigrade;* this would be the same as $\frac{1}{490}$ for one degree *Fahrenheit.*

† The temperature of melting ice is 32° Fahrenheit, and 0° Centigrade.

and we have supposed that we were heating it one degree, while allowing it to expand in such a manner that the pressure should remain constant; we have afterwards supposed that we were heating it one degree, while preventing it from changing its volume. Will there be necessary in each case, in order to produce this same elevation of temperature, a like quantity of heat? Evidently not. Under a constant volume the air has no outside work to accomplish. Under a constant pressure it must displace the exterior obstacle which opposes its expansion; it has thus a real work to accomplish. In this second case it must absorb an excess of heat which is exactly the equivalent of the work done. The heat-capacity under a constant volume, and the heat capacity under a constant pressure, differ then in a notable manner. For air, they are in the ratio of 1 to 1.421. The difference between these two quantities represents what was formerly called the latent heat of expansion, and what is now the precise equivalent of the work which the air must do in order to become expanded.

We may even observe that it was from this expansion of the air, the numerical conditions of which have for a long time been fixed, that Dr. Mayer sought to obtain, in 1842, a primary determination of the mechanical equivalent of heat. The number which

M. Mayer deduced does not differ sensibly from that which has been definitively adopted after a series of experiments of every kind.

We have said that air accomplishes outside work in expanding; such is usually the case; but it may, under particular circumstances, expand without having performed any work. Now it is the work which absorbs the heat, and not the expansion itself; if there is no work in the expansion, it is not indicated by the absorption of heat.

This phenomenon has, moreover, been proved by a celebrated experiment which M. Joule made in 1845. M. Joule took two receivers of the same size, connected by tube and stopcock; into one he put air under a pressure of twenty-two atmospheres, in the other he made a vacuum, and he permitted the compressed gas of the first receiver to expand in the second, and a state of equilibrium soon ensued under a uniform pressure of eleven atmospheres. In order to arrive at this state, the gas had not had any outside work to do, and M. Joule showed that the temperature was the same at the beginning as at the end of the experiment. Doubtless there were, at certain moments, changes of temperature, but the partial losses and gains compensated each other, and at the final analysis the absence of work done was indicated by the absence of variation in the temperature.

The experiment of M. Joule has been repeated by several scientific men, and in a noteworthy manner by M. Victor Regnault. But it demands a high degree of precision, and is not of a nature to be reproduced in a lecture upon physics. Mr. Tyndall, in his course at the Royal Institution, shows the results of it by means of suitable and familiar apparatus.

He first takes a box in which a certain quantity of air is compressed, and he opens its stop-cock to let the gas escape. Here the gas no longer finds a vacuum before it; it must, in order to expand, drive away the external air; it must perform a work; and it is only from itself that it derives the necessary heat. There is, then, a lowering of temperature, and Mr. Tyndall renders this result visible by directing the jet upon the face of a very sensitive thermo-electric pile;* the needle of the galvanometer indicates the cooling of the gaseous jet. Instead of the box of compressed air, Mr. Tyndall next takes an ordinary

* Mr. Tyndall has a thermo-electric pile so sensitive that, maintained at a temperature of about fifty or sixty degrees, it indicates at a distance of twenty paces the heat which is given out by a man's body.

pair of bellows, and in working them he directs the jet upon the front of the pile. In this case the gas itself has not to yield the heat necessary for pushing away the external air; the hand of the operator furnishes the work directly; it even furnishes it in excess; and the needle of the galvanometer, instead of indicating a diminution, marks an elevation of temperature.

The theory of heat is becoming more complete every day; but even now it is sufficiently advanced to present a respectable state of entirety. If it still exhibits gaps and doubtful points, the principal outlines are at least clearly marked. The molecular motions which constitute heat are not immediately perceptible to our senses, but they may be said to lack but little of it. They are even almost discernible, so well known and so exact are their mechanical effects. When the living force passes from the molecules to the mass of a body, and returns from that mass to the molecules, thus appearing successively under the form of work and of heat, we cannot observe these changes; but the phenomena, a little before and shortly after the transformation, are so well determined, that we believe we see the change itself.

Thermo-dynamics is a field sufficiently explored;

one in which the mistakes of the way are not serious, and in which one is certain, having gone astray, to regain his path. Investigating electrical phenomena, we are about to enter a region far more obscure and dangerous.

CHAPTER IV.

ELECTRICITY.

I.

It is necessary to determine the Electric Unit, and to find its Mechanical Equivalent.

WHAT is electricity? How shall we regard that common conception which is based upon the play of a positive and a negative fluid? Are there, in reality, two electric fluids? Is there even one?

We put these questions; but a reference to the premises already laid down will leave but little doubt as to the kind of answers we shall give them.

And first, the duality of the fluids can no longer be regarded other than as a figurative expression. We may even ask ourselves if it ever had an appearance of reality. It has all the characteristics of a fiction of analysis; it carries the mind at once into the realms of mechanics. In mechanics, motions are termed positive or negative, according to their taking place in one direction or another; so the hypothesis of a duality of fluids becomes resolved into a mathematical conception.

Is there even any special fluid to which we must attribute electrical properties? It would doubtless seem proper to make here at the outset, some partial reply, and not to decide this question without some reserve; but we need not again declare that we have banished all idle prudence, and we do not hesitate to assign, at once, to the electric fluid a place outside of science, and send it to join the company of those delusions of the past, — the calorific and luminous fluids and the many so-called entities of former times.

With regard to magnetism, we may leave it entirely aside, since standard instruction has long ago traced to one and the same principle both the magnetic and the electrical phenomena; a permanent or a temporary magnet may be regarded as the seat of a series of little currents setting in the same direction.

The field being now cleared, the question presents itself to us in this form: Is electricity a motion of the ether? Is it a motion of ponderable matter? Is it a motion of both? In a word, what is the character of this motion?

Before broaching these questions, we desire to call attention to two important and decisive points in the study of electricity.

Electrical phenomena have been studied with great care during the past few years; a vast number of little facts have been collected, which present, however, only a confused appearance, being badly grouped, and throwing but little light upon each other. This is doubtless due in a measure to the nature of the subject; but it is also attributable, in part, to those who make the observations. One essential and primary condition is wanting in the researches which have been made here and there on the subject of electricity, namely, an agreement as to the unit to which all these actions shall be referred.

We have already had occasion to mention the capital importance that attaches, in physics, to the choice of units. Every phenomenon results from the coexistence of a certain number of correlated facts, and in order to illustrate the relation of these facts, it is necessary to represent each of them in its proper quantity by a particular variable. Thus, if we try to define the orbit of a planet about the sun, we must take for the elements of our research, on the one hand, the variable length of the radius vector which joins the sun to the planet, and on the other, the constantly changing inclination of this radius to the axis of the perihelion; observation will show forthwith the relation between these two quantities which constitutes the equation of the ellipse, and we may declare that the planet runs in an elliptical orbit, in which the sun occupies one of the foci. It would not do, however, to suppose that a phenomenon would be equally easy to define with any variables whatever that might be selected; on the contrary, this selection exerts a very decisive influence upon the results obtained. With certain variables you will arrive at only confused results, from which you can derive no profit, while with others, we shall bring to light precise laws.

We might thus mention, in the history of physical science, many an unfortunate selection which has retarded important discoveries, likewise also many fortunate guesses. An example of the latter occurred in the case of Kepler's first law, the second of which we just now alluded to. When Kepler sought the law of the motion of a planet in its orbit, he selected as variables the time, and the areas described by the radius vector. It would have been just as natural, more so perhaps, to seek a relation between the time and one of the variables mentioned above, namely, the length of the radius, or its inclination to the line of the apsides. Had Kepler pursued this course, he would not have found any simple relation between the numerical values resulting from his observations and those of Tycho-Brahé; the intimate connection of these values would have been concealed under relations so complicated that it never could have been demonstrated. On the contrary, thanks to the variables he had chosen, Kepler was readily able to observe that the numerical values representing the times, and those representing the areas, formed two proportional series. In this way became plainly evident that great law of astronomy, which we express by saying that the areas described by the planets are in proportion to the times, or that planets describe equal areas in equal times.

A fortunate selection of variables is, then, indeed an essential condition of success, and constitutes almost the chief difficulty in all physical researches. How much more important becomes this consideration when we are treating, not of quantities which serve to verify a particular law, but of those which form the standard of a whole class of phenomena.

We now see the first step to be taken by electricians. They must agree upon some common and convenient measure of electrical actions. Failing in this agreement, they work each for himself, unable to systematize their discoveries, and never arriving at a mutual understanding; there reigns among them a confusion of tongues.

Who will put an end to this ? Who will furnish a basis of common agreement ?

Five years ago the British Association made laudable efforts in this direction. The British Association is, as is well known, a private society, devoting itself, in England, to the advancement of science, and whose watchful attention is directed successively to all points where there is urgent need of making investigations. To aid the progress of the submarine telegraph, it appointed, in 1862, a commission, which examined the whole question of the measure of electrical phenomena, and proposed a solution strictly applicable, though very complicated.*

* A previous commission had been instituted in 1861. Its special object was to determine a scale of *resistance* by which to test the value of submarine cables, manufactured in English workshops, with respect to their transmitting qualities. The labors of the British Association have exerted no small influence on the wonderful improvements which have been brought about in the manufacture of cables in England, which have at last resulted in establishing between Europe and America telegraphic communication. The commission of 1861 was succeeded, in 1862, by a new commission, consisting of Messrs. Wheatstone, Thompson, C. W. Siemens, and Charles Bright. This new commission has not limited its work to the measurement of resistances; it has confronted the whole question of electric units, seeking to connect them closely with the units employed in mechanics. Experiments have been made at King's College to determine what degree of precision is practically attainable through the application

In France this problem would not even seem to be the order of the day. We have, it is true, an association for the advancement of the physical science of the globe; but its members would seem to have nothing left to desire, while they have the moon pointed out to them every month at the observatory.

Meanwhile, be the question of electric units decided on this side of the channel or the other, the study of heat clearly indicates the character of the solution which must eventually result. So long as calorific effects were estimated merely by changes in the thermometer, we remained on the outside of phenomena, and knew nothing of their essential nature. Temperature is only one of the peculiarities of heat. I have a kilogramme of water at one hundred degrees Centigrade. If it evaporates freely in the air, it absorbs the enormous quantity of 536 heat units, and the kilogramme of vapor which results is still at one hundred degrees.*

of the theoretical views of the commission. The result of these investigations is contained in a report drawn up by Mr. Fleeming Jenken, and published by the commission in the form of an appeal to the scientific world.

* Sometimes a very elegant experiment is made in the laboratory, showing that different bodies, while at the same temperature, contain very different quantities of heat. A cake of bees-

Between the motions which take place in the interior of bodies and the variations they effect upon the thermometric scale, there exist only indirect, and, so to speak, accidental relations. The study of these relations has never afforded more than vague and confused information. Real progress began the day when calorific phenomena were no longer referred to the degrees of the thermometer only, but to an intrinsic unit, —the heat unit; that is to say, the entire quantity of heat necessary to effect a certain definite result, and one easy to appreciate.

Hitherto the galvanometer has been almost exclusively employed as the measure of electrical phenomena. Now, we may remark in passing, that the galvanometer is a far more imperfect instrument than

wax, about twelve millimetres in thickness, is suspended from a support; next a vessel of boiling oil is taken, and balls of different metals of the same size are plunged into it — balls of iron, copper, tin, lead, and bismuth, for example. These balls, having all taken the same temperature, that of the boiling liquid, are taken out of the oil, and placed all at once upon the cake of wax. They sink into the wax, but with different degrees of rapidity. The iron and the copper enter powerfully into the melting mass; the tin more gently; the lead and bismuth remain behind. The iron ball passes through the wax, and falls out first; the copper one follows it; the others remain in it, unable to pierce the cake, and stop there at different depths, according to the laws of their calorific capacity.

the thermometer even. The thermometer, at least, indicates directly by its linear expansions that part of the calorific matter it is required to exhibit. The galvanometer, which also indicates but a part only of the electrical effect, has the further disadvantage of exhibiting them only by the angular variation of a needle. We are, then, obliged to compare angles, that is, to estimate sines and tangents. Already excluded from actual contact with the facts, the observer finds them still further masked by trigonometrical functions.

There is, then, urgent need of penetrating to the very core of phenomena. In all our succeeding investigations, we must take for our fundamental idea the electric unit, that is, the amount of electricity required to produce a fixed result.

What shall be the effect henceforward selected for our type? Here is a question admitting of discussion. Suppose, just to fix our ideas, we select the voltaic decomposition of a kilogramme of water. The electric unit thus determined, we shall be compelled to express, by means of this fundamental unit, the various electrical phenomena that have hitherto been characterized only by special circumstances, by the intensity of the current, or by the amount of heat developed. Instead of halting at partial effects, we shall approach the facts as a whole. Out of the mass of incoherent observations now presented by the science of electricity there will arise a sort of natural selection; isolated laws will be gathered into groups, and their inner meaning made manifest.

To choose the electric unit,—this is the first step in advance to be made by electricians; the second is to ascertain the mechanical equivalent of electricity; to find out how many kilogrammetres are equivalent to an electric unit.

We now see, by a characteristic example, the utility of a hypothesis capable of comprehending natural phenomena as a whole, and of tracing them back to a single principle. By it the natural philosopher may be guided in the imperfectly known regions he explores; by it instructed in the path he must pursue through the labyrinths of particular facts.

Let us observe, however, that in order to take the two steps we have mentioned, it is not necessary to get a preliminary view of the nature itself of electricity. If we look into the history of heat, we shall see that the notion of a heat unit was not at all peculiar to those who regarded heat as a motion. It might even be remarked that this unity has a suspicious look, and that it savors a little of the doctrine of the materiality of caloric. The equivalence of heat and mechanical work has also been established outside of all theory. This notion of equivalence is a prudent and eclectic one; it involves no preconceived idea as to the facts one is comparing; they are equivalent, nothing more. When one is sure that he is comparing together two motions, the words equivalent and equivalence become, so to speak, inadequate, and he has the right to resort to more energetic terms.

First, to decide upon the electric unit, and then to determine its mechanical equivalent, such are the two points to which the efforts of electricians must first be directed, and the ones we have desired to bring to view. Having given these general suggestions, it remains for us to show what experience teaches us at present with regard to the conditions which characterize the electric motion.

II.

The Electric Current apparently a Transport of Ethereal Matter.

THE preliminaries just laid down show clearly enough that we are far from possessing a general theory in regard to electrical phenomena.

We were not in want of experimental data. Observers have placed at our disposal a large number of facts, too many, almost, since the special laws established by them are not referable to a few principal groups; they exhibit only one phase of each phenomenon, and have, for the most part, but an obscure and commonplace significance. Nevertheless, from a general view of these confused observations, we conclude that electric motion is a real transfer of matter; the word *current*, employed in ordinary language, would thus correspond to the real nature of the phenomena.

A decisive consideration may be brought to the support of this opinion. If the two poles of a battery are connected by a conductor of variable dimensions, the intensity of the current, as measured by its effects on the galvanometer, is the same in every part of this conductor; wherever it becomes thinner, there the current is more rapid, so that each segment gives passage in the same period of time to the same amount of electricity. This peculiarity is easily made visible by its calorific or luminous effects. We know that if a very fine wire is interposed in the passage of a current it grows red, and becomes heated even so far as to melt. We are also acquainted with the experiments made with Geissler's tubes. These are glass tubes, in which the air is rarefied, and which are laid in the course of the current, in order that the electricity may traverse them in the form of a luminous spray. Now, if we take a Geissler's tube, varying in size in its several sections, we may easily demonstrate that the spray becomes the more luminous the narrower the tube. In the fact that the motion increases in proportion as the calibre of the tube diminishes, we recognize a fundamental law of the flow of liquids; a law known since the time of Leonardo di Vinci. This fact alone excludes the idea that electricity may be the result of simple vibrations. It does not appear, in fact, in any of the vibratory motions we are acquainted with, whether longitudinal, like those of sound, or transversal, like those of light.

When these latter motions encounter an obstacle which contracts the medium in which they are displayed, they are reflected into the body of the medium; but they do not hurry forward into the open space before them; it is those fluids which are endowed with a motion of transportation that are thus accelerated in narrow passages. When an iron rod is heated, we do not observe the temperature to be any higher in those parts where the rod is thin. The case is otherwise when the elevation of temperature is caused by electricity, since, as was just now stated, very fine wires placed in the circuit of an ordinary conductor may be heated to a red heat and melted.

We find, then, at the outset, by means of a fundamental fact, that the electric motion is similar to the flowing of a fluid. This analogy may be traced through all the particulars revealed by experiment.

The science of telegraphy, especially, has furnished us numerous hints in this direction. A telegraphic wire is like a tube which is to be filled; the battery is like a reservoir which fills the tube more or less easily, according to the degree in which it is itself filled. Be the wire charged, or half charged, when the end which communicates with the battery is placed in the ground, a part of the charge flows back. The case is similar with a liquid flowing from a tube open at both ends. Nothing of the kind would have taken place in the case of a vibratory motion; such a motion, when the cause producing it has ceased, does not take a backward direction, but continues precisely in that in which it began.

By such examples we are guided in considering the duration of the propagation of a current, that is, the time necessary for the current to attain a uniform state throughout the whole extent of the wire; and here again we are reminded of the transport of a fluid, since the time increases nearly in proportion to the square of the length of the wire. This time varies inversely as the diameter of the wire, and this fact alone shows that we have not to do with a vibration; a vibratory motion, in fact, assumes its uniform condition as rapidly in a large tube as in a small one. This may be verified in the case of sound.

But what is this fluid, the transportation of which constitutes the electric current? Is it, perchance, ponderable matter itself, reduced to a state of vapor, or at least brought to a condition of tenuity, which imparts to it the properties of fluids? Certainly not. For, in the first place, we have no reason to suppose that the passage of a fluid through a wire augments its weight; besides, if the electric flux were a transport of ponderable matter, if the matter itself of the conductors were transported, it ought to be perceptible when two unlike wires are joined to each other; when the current, after passing through a copper wire, for example, passes into an iron one, the copper ought to leave some traces of its passage through masses of the iron, and *vice versa*. Observation has not disclosed any fact of the kind, unless it be at the very point of junction. Even here, we must admit, the transportation of matter is a mere accident, an accessory phenemenon, a purely local circumstance, that may be neglected without hesitation.

Are not our conclusions, then, self-established? This fluid, which is carried along a conductor, is nothing else than the imponderable matter we are acquainted with under the name of ether. The electric motion of the ether is, moreover, in no sense a vibratory one ; it is a veritable flux, an actual transport.

We cannot but be confirmed in these opinions if we make a further hasty examination of some of the peculiarities presented by these currents.

The electric spark has been thoroughly investigated. It offers an interesting subject for study. Physicists have always hoped to find here, under a striking form, some direct information concerning the nature of electricity. They have especially studied the spark proceeding from static machines ; but their conclusions might legitimately be extended to that produced by currents.

It is necessary to state that the study of the spark has long been productive of deceptive arguments, and especially has it been of service to the theory of two fluids. On beholding the spark, compact and brilliant at both poles, but larger and dimmer at its centre, one was sure that he saw the two different fluids in the very act of combination. There was the positive fluid proceeding from one pole in the form of a fan, and the negative fluid escaping from the other in the form of a star. To us the brilliancy of the two poles seemed to proceed from the agitation produced in them by the electric flow ; but the flow may likewise produce this effect when escaping from one side and entering the other.

Nevertheless, to prove that a fluid passed out of both sides at the same time, an experiment was made, which seemed decisive. The spark was made to pierce several sheets of paper, and it was shown that the edges of the aperture were turned, some towards the negative pole, some towards the positive pole, — a fallacious result, and one from which no conclusion can be drawn as to the direction of the electric current at either pole. In many cases a body forcibly punctured exhibits edges turned in a contrary direction to that of the pressure. It would look, therefore, as if the penetrating body has, in the second stage of the perforation, experienced a rebound. The symmetry of the ridge raised on both faces of the sheets pierced by the spark affords, therefore, no proof of the passage of a double fluid.

On the contrary, the recent advancements in spectroscopy go to prove the unity of the motion. It has been proved that the spectrum of the spark depends upon the nature of the metal composing the positive pole, since it remains unchanged when the nature of the other pole is altered ; the metallic particles carried along by the current show, then, that the transfer takes place in a single direction.

Another important fact is, that the spark is stratified. It is seen in layers ranged one upon the other. It would appear, in fact, as if the electric flow was not a continuous one. We have here developed a phenomenon similar to that which is produced when we see smoke issuing from a chimney in successive puffs. When a flowing body meets an obstacle, it produces, in the effort to overcome that obstacle, certain onward movements which arrange themselves over each other. May not also the stratification of the electric spark be like the transfer of metallic particles — a purely local accident? May it not indicate a state of things existing throughout the whole extent of the conductor? We might assert, then, that the transportation of the ether is effected by successive waves; but these waves, which accompany a movement of transportation, must not at all be confounded with vibratory waves, of which light and sound are examples.

It will be seen that we make here an important reservation. We have thus far admitted that the ether is really carried from one end to the other of the conductor; that each atom employed in the circuit runs through its entire length. It is possible, on the other hand, that each atom is permitted only a partial range, and that the current is produced somehow by a series of relays more or less near together. We leave the door wide open for such a supposition, which is in no way incompatible with what we have above stated; but, for simplicity's sake, we shall continue to speak as if we were treating of a fluid in motion, the particles of which all pass through the entire circuit.

One other question presents itself in the study of the spark, — a question of high importance, and one which can be only partially answered. Is the spark produced in a perfect vacuum? In other words, can the electric stream, even though it be itself nothing but a motion of ether, exist outside of ponderable matter? The importance of this question is manifest, and it finds no answer in the general phenomena of nature. The sun sends us light; we receive electricity from it, not directly but mediately. The frequent experiments made to discover whether the electric spark can pass through an absolute void, have been subject to much dispute. How shall we procure an absolute void? We try to empty a tube of all ponderable matter; we fill it several times with carbonic acid, which we expel by means of an air-pump, and finally we use potash to absorb the remains of the acid; but are there not vapors escaping from the joints, from the valves of the machine, and from the potash itself? How get rid of this source of error? Therefore, we repeat, nothing could be less satisfactory than the result of such an experiment. The attempts which have been made, however, go to prove that the spark does not pass through a vacuum, and from considerations of quite another kind we are led to the same conclusion. It would seem, then, that the electric movement can only be produced in the midst of ponderable matter.*

Let us now direct our attention to those phenomena in which currents have their rise, the two principal ones being heat and chemical action. How shall we conceive, in either case, of the beginning of a current? If two metallic bars, one of bismuth and one of antimony, for instance, are soldered at one end, and we heat the point of union, a current arises in the arc connecting the two metals. Such is the principle of the thermo-electric pile. Observe that we must have, at the place where the heat is applied, different metals; a junction of different sections of the same conductor would not suffice to produce a current; we must have

* This statement seems to be confirmed by experiments made with Plücher's tubes, used in spectrum analysis. In a tube, reduced as nearly as possible to a vacuum, no spark is observed, although the platinum wires on the ends of the tube are very closely approximated. In a second tube, containing the slightest possible quantity of hydrogen, the light of the passing spark is clearly visible. (See Schellen's Spectrum Analysis, p. 26.) — *Translator.*

molecules which are unlike; and what does this signify? Let us refer again to the hypothesis we have constructed to explain how bodies pass from a gaseous to a liquid and solid state. We were compelled to admit that every molecule carries along with it in its rotation a sort of atmosphere of ether. When unlike molecules are placed side by side, then there will be a meeting of atmospheres of different densities and velocities; and if their equilibrium becomes disturbed by the application of heat, we may see how this circumstance would set free a certain number of ethereal atoms. These atoms rush into the conductor as into a channel, and there form the current. The more discordant are the atmospheres of the two metallic elements, the more intense will be the effect; there will be no effect when the atmospheres are all alike, that is, when only one metal is employed. Chemical action produces an analogous effect on a larger scale. When two bodies combine, the molecular atmospheres are powerfully disturbed; a new distribution of the ether is forcibly effected about the new molecules, and this sudden change drives away more or less of the ethereal atoms. Thus different batteries, the thermo-electric pile as well as those based on a chemical combination, exhibit at the very origin of the current the beginning of a flow of ether.

Beginning in the pile, this flow is continued in the conductor, and if we regard the entire circuit thus formed, we shall readily see that chemical action, electricity, heat, mechanical work, are all produced according to that law of mutual transformation to which we are compelled to reduce all physical phenomena.

The *vis viva* due to the action of the pile sets the ether in motion; this, circulating in the conductor, develops in it heat, because it agitates in its passage the ponderable molecules, and leaves them a part of its *vis viva.* But instead of producing heat, it may produce work of a different kind.

We shall find a ready example of this by placing in the circuit a voltameter,* filled with water. The two poles of the current, the two electrodes of platinum being directed into the upper part of the liquid, the water becomes warm, and soon boils; then if the poles be more deeply plunged into the vessel, the water begins to resolve itself into its two elements, the temperature of the liquid diminishes, and we observe very soon the ordinary conditions of electrolytic decomposi-

* An instrument to measure the strength of an electric current, consisting of a graduated tube, which receives and measures the amount of gas generated by the current in a given time. — *Translator.*

tion which are accompanied by a slight elevation of temperature.

We see here an electrolytic and a calorific action directly exchanging with each other. If the experiment was so conducted as to yield precise measurements, if we were able to free it from every source of error, we should discover just what weight of water may be heated one degree by the quantity of electricity which will decompose a given weight of that water; in other words, we should find the relation between the electric unit and the heat unit, and electric currents would thus be reduced to the common measure of mechanical work, to the kilogrammetre.*

In the example we have given, the current produces a chemical work; it might also produce a mechanical

* Father Secchi has made some experiments, from which we might conclude that the quantity of electricity which decomposes 0.106 gramme of water, would raise, by one degree, the temperature of thirty-eight grammes of the same liquid. Taking for our electric unit, as we have above suggested, the quantity of electricity capable of decomposing a kilogramme of water, it will be found that an electric unit is equal to three hundred and sixty heat units, or one hundred and fifty-three thousand kilogrammetres. If, for the sake of smaller numbers, we compare the electric unit with the gramme, it will then be equivalent to 0.36 heat units, or one hundred and fifty-three kilogrammetres. We give this result, but we are not willing to certify that the experiment from which it is derived can be regarded as covering all the conditions of the problem.

work, — raise a weight, or turn a shaft. M. Favre, in his series of well-known experiments, has shown that the heat developed in a current decreases precisely in proportion to the work produced. The *vis viva* of the electric current is in part consumed by the lifting of the weight or the turning of the shaft, and the calorific disturbance of the circuit is diminished in proportion. We see electricity converted into work, instead of being transformed into heat; if this conversion could be complete, if we were able to eliminate entirely from the experiment the manifestation of heat, we might at length determine exactly the ratio of equivalence between electricity and mechanical work; we might observe directly the relation between the electric unit and the kilogrammetre.

But this is a purely theoretical conception; if, as is probable, the electric flow takes place only through ponderable matter, it necessarily sets in motion its molecules; that is to say, there is no electricity without heat. We must mention in this connection, that the conductibility of different substances follows almost the same order, both for electricity and for heat. If, for example, we regard the metals in this twofold relation, not only do they arrange themselves in both cases in the same order (silver, copper, gold, tin, iron, lead, platinum, bismuth), but the same series of figures might represent exactly their double conductibility. The close connection existing between calorific and electric phenomena, hardly permits us to hope that the mechanical action of electricity may be isolated in practice, or reached by direct observation.

III.

Electricity and Light; their Mutual Relation.

In the degree that we have examined the peculiarities which mark the propagation of the currents, the origin of the electro-motor forces, and the distribution of work in the conductors, we have become confirmed in the idea that electricity consists in a transport of the ethereal fluid, of the same fluid that produces light.

This will, indeed, be to many minds a resemblance hitherto unexpected. To compare light and electricity is quite a new idea, and yet we have just been regarding both as different motions of the same fluid. Between these two modes of motion a new bond appears.

If we look at the generality of natural objects, we shall notice that those which are transparent are usually non-conductors; permeable to light, they refuse passage to electricity. On the other hand, conductors are generally opaque; witness all the metals.

The objection may, perhaps, be urged that water is both transparent and a conductor; gutta-percha, opaque and a non-conductor; but let us consider here extreme cases only, neglecting those of an intermediate character. We see, then, two clearly marked general groups, transparent bodies and conductors. These are ill-chosen designations, since they convey no idea of contrast; but beneath the terms let us look for the facts. Among bodies of the first class, the ether moves transversely only; on the other hand, it can take a longitudinal motion only in bodies of the second class. Difference of molecular aggregation creates, therefore, a difference of mobility with regard to ether. This is all we can say; but we may assert that the two classes of bodies enclose not two different fluids, but one and the same ether, susceptible of a variety of motion.

To admit the existence of a luminous fluid belonging to transparent bodies, and of an electric fluid peculiar to conducting bodies, would lead to very strange results.

When lead combines with silica to make glass, we should have to suppose that the electric fluid is driven away and replaced by the luminous fluid! The diamond in becoming charcoal, is no longer transparent and non-conductor; it becomes opaque and a conductor. Here, then, would be another change of fluids. Such a thing cannot be conceived as possible. We can, on the contrary, very readily imagine how a single ethereal fluid, following the molecular arrangement of bodies, may sometimes find its motion obstructed in one direction, sometimes in the other.

We will here add an argument furnished by the velocities with which light and electricity are propagated. The velocity of light is about one hundred and eighty-five thousand miles a second. That of electricity has been determined with far less exactness, since it depends upon the nature of the conductors, and a variety of circumstances which observers have not succeeded in eliminating. But taking a mean between the widely different results yielded by experiments, we shall not come far from the same rate of one hundred and eighty-five thousand miles a second.

In this resemblance may be seen a confirmation of our hypothesis. It should not in the least surprise us that the same number represents two velocities corresponding, in our view, with two motions of the same fluid in the same direction; the velocity of light is, in fact, that of the longitudinal impulse from which the transversal motions result.

Such are but very general views of this subject, and it can hardly be said that we see clearly the connection between the phenomena of light and of electricity. Scarcely can we conceive of the conditions capable of effecting this twofold faculty of motion in the ether.

We are acquainted with the ingenious explanations Father Secchi had recourse to in order to show how the impulse sent along a luminous ray betrays itself in transverse vibrations. Other hypotheses have been proposed to show how these transversal vibrations can be extinguished in conducting bodies to the advantage of the longitudinal ones. But let us leave these problems in their obscurity, — it is quite fitting that we should conclude our hasty view of the phenomena of electricity with an avowal of the uncertainty of our position. These phenomena still offer much that is obscure, and it is only in keeping with the actual state of our knowledge that we dismiss them, leaving important questions still pending.

CHAPTER V.

THE ATTRACTIVE FORCES.

I.

Points of Resemblance and of Dissimilarity presented by Gravity, Cohesion, and Chemical Affinity.

THE charcoal points of an electric lamp grow hot, and become luminous, when a current passes through them. A furnace fire gives out heat and work. A ray of light falling upon a sensitive plate determines an electro-motor action, which becomes motion in the needle of a galvanometer, heat in the thermometric index. We might multiply to infinity similar examples in which light, heat, and electricity appear as convertible phenomena, or reducible to the idea of mechanical work. Work produces them, and they produce work. They originate in motion, and they are resolvable into motion. The public mind is accustomed — so, at least, it seems to us — to regard in this manner the effects of light, heat, and electricity; but the point is not so well settled in regard to the attractive forces, gravity, cohesion, and affinity, which appear to reside in the recesses of matter. Until now they preserve a more mysterious aspect. It remains for us to see if we shall be able to dissipate, in part, the obscurity which surrounds them by applying the principle which now enables us to throw light on all natural phenomena.

In the progress of our work we have briefly indicated, as opportunity offered, the considerations which enable us to reduce these forces to the effects of motion. We have especially, then, to classify and develop here the hypotheses previously broached.

In the first place, the attractive forces should not be considered as inherent in matter.

When Newton proclaimed the law of universal gravitation, he took good care to make his reservations in regard to it. After having described the planetary motions in his book, The Mathematical Principles of Natural Philosophy, he adds, "I have thus far explained celestial phenomena and those of the sea by the force of gravity; but I have nowhere assigned the cause of this gravitation. This force comes from some cause which penetrates to the very centre of the sun and planets, without losing any of its activity; it acts according to the quantity of matter, and its action extends in all directions to immense distances, always decreasing in the inverse ratio of the square of the distances. I have not yet been able to deduce from the phenomena the reason of these properties of gravity, and I do not conceive of any hypothesis. . . . It is enough that gravity exists; that it acts according to laws that have just been exposed; and that it can explain all the motions of the heavenly bodies, and those of the sea." Again, in the same book he says, "I mean by the word Attraction, the endeavor which bodies make to approach each other; whether this endeavor results from the action of bodies which mutually seek each other, or which influence each other by means of emanations, or whether it result from the action of the ether, of the air, or any other medium, corporeal or incorporeal, which urge towards each other, in some way or other, all the bodies which float in them."

Thus Newton left this question undecided; but after him it became gradually a custom to consider gravity a kind of quality inherent in bodies. Many people admit to-day, as a primary axiom, that matter is inert, and, for the second, that it attracts according to such and such laws. We have already said that it is necessary to choose between these two contradictory ideas. If the molecules are drawn towards each other by virtue of a cause which is within themselves, how can you say that they are inert? They are active, on the contrary, and all the structure which you have raised upon the idea of inertia crumbles to its foundation.

What will be the case, then, if we pass from gravity to chemical affinity! If the molecules exercise a choice by virtue of an inherent principle, they have, then, a primary, active principle of their own; they have wills, caprices! Chemistry becomes the study of the molecular passions. We shall find in it sympathies and antipathies; base instincts and noble sentiments; lawful affections and culpable desires; happy marriages and ill-assorted unions; half-concealed quarrels and open contests. Such are the idyls and the dramas which chemistry presents to us if we place in the molecules a repulsive and an attractive principle, as formerly a spirit of good and a spirit of evil were made to dwell in human souls.

It is a pure geometrical fiction to suppose that two molecules act upon each other at a distance. In reality we are only acquainted with actions which take place by contact, by the communication of motion. Between the molecules are the ethereal atoms; shocks are transmitted from one to the other; the matter remains inert, and is only excited to motion on the side where it is struck. The repellent forces have already disappeared before the idea of calorific motion; the attractive forces must likewise be reduced to the effects of impulse.

When we compare the three forces, which we find grouped in the same family, — gravity, cohesion, and chemical affinity, — we are at once struck with the disproportion between them.

How much more powerful is cohesion than gravity; an iron wire will not break under its own weight until it reach a length of five thousand metres. Enormous masses of metal are needed to overcome by their weight the cohesion which exists in a single section of the wire.

What is more extraordinary still is, that when once the adhesion is overcome, and the wire broken, the closest contact of the disjointed parts reproduces no trace of the primitive cohesion. Thus cohesion, incomparably more intense than weight, is sensible only at extremely small distances; weight, more feeble, on the contrary, continues its action at infinite distances.

If any one is desirous of getting a comparative idea of these different forces, he may have recourse to the following hints. They are due to a learned physicist, M. Dupré, who for long years has devoted himself to the study of molecular activities.

M. Dupré deduced from his experiments and calculations the force necessary to overcome the mutual affinity of the elements of water; to separate by force oxygen and hydrogen over the surface of a square millimetre. He found this force would be about sixteen hundred and seventy-three kilogrammes.

To overcome the molecular adhesion of water, to tear away one layer from its neighbor, a force would be required of seventy kilogrammes to the square millimetre.*

Finally, it is known that over this same surface gravity exerts an action of only 10.33 grammes.

Comparing the three numbers which represent the power, respectively, of affinity, cohesion, and gravity, we can appreciate the enormous difference in their values.

II.

Gravity may be considered as an Effect of the Motions of Ether and of Ponderable Matter.

LET us attack, without further delay, the theoretical considerations which may enlighten us as to the nature of the attractive forces, and let us begin with gravity.

Let us imagine the ether uniformly diffused throughout space. Its atoms, endowed with motions of pro-

* That is, nearly fifty tons to the square inch! — *Translator*

gression and rotation, strike each other in the manner already mentioned.

Let us now suppose that, at some point within, there is a special and permanent disturbing cause, as, for example, a molecule having weight, and itself endowed with a vibratory motion. The shock goes on, extending throughout the ethereal mass, and by reason of the nature of this medium is propagated in all directions. The atoms nearest to the heavy molecule will receive violent shocks ; they will be powerfully urged, and their ranks will grow thin in the neighborhood of the centre of disturbance, and the layer contiguous to the molecule will become less dense than the rest of the medium. The motor action continuing, this same effect becomes propagated from layer to layer throughout space. As a final result, the ether becomes arranged around the centre of disturbance in concentric layers, the first of which, and nearest to the molecule, will be least dense, and they will go on indefinitely increasing in density. This condition of things might be easily represented and the figure traced ; the molecule at the centre, around it spheres of atoms, wide apart at first, then nearer and nearer to each other. Let us remark in passing, that the difference in density of contiguous layers, like all effects which are propagated by concentric spheres, is inversely proportional to the surface of these spheres, that is, to the square of their radii.

This established, suppose a second molecule to be situated at any point of this system. It will encounter on the side towards the first molecular layers of ether less dense than upon the opposite side ; pressed upon by the ether in all directions, it will receive, notwithstanding, fewer shocks on the side towards the first molecule, and it will consequently tend to move towards it.

Such would seem to be the cause of gravity.

The second molecule is pushed towards the first, because it encounters ethereal layers of different densities, and the energy of this action, for the reason we have just now pointed out, is inversely proportional to the square of the distance between the two molecules. In this statement we recognize the law by which gravity acts.

What we have just said concerning isolated molecules, is also applicable to those grouped in a way to form a body. Such a group will effect in the ether that variation of density we have described ; it will effect it with so much the more force as the molecules are more numerous, or the mass of the body greater. The stars, in fact, are only huge bodies, impelled by the same cause that makes heavy subsfances fall to the surface of the earth. With the former, as with the latter, the attraction is only that tendency to approach, the origin of which we just now referred to external impulses.

Of course the brief and general outlines just given do not constitute an exact demonstration. To throw light upon a question of such high importance, it would be necessary to pursue the phenomena into their minutiæ; to exhibit in detail the various rebounds made by the ether, which result in its arrangement about the molecules in layers of different density. It would be necessary to anticipate the doubts generated by such an exhibit, and to reply to the principal objections that might be offered.

For example, it might be asked why the effect we describe is peculiar to the material molecules; why it is not produced, at least here and there, around the ethereal atoms. The answer to this is easy. In the midst of the ethereal mass, in the absence of any molecule, everything is symmetrical with regard to each atom; the effect begins, if you please, around each of the atoms; it is as if it did not begin around any of them, and the medium remains uniformly dense; to break its uniformity, there is needed a centre of disturbance.

It will be asked, again, if it is not a very arbitrary

supposition to give to atoms, and especially to molecules, the round form which seems necessary at the start, in order to explain the regularity of the shocks and the symmetry of their effects. Here again the answer is easy. The theory of rotation teaches, in fact, that the shocks do not depend upon the exterior form of the bodies, and that we may always conceive of a solid, of any form whatever, as being replaced by an ellipsoidal globe. The round form is not then really necessary, either to the molecules or even to the atoms.

There would be many other objections to overcome; but it is easy to see that we could not here analyze all the circumstances of the phenomenon. Our end is attained if the general principle of the explanation just given has been grasped, and if it is seen how the motion of the ether may produce terrestrial as well as sidereal attraction.

There remains a point, however, upon which we cannot help saying a few words.

It may be observed that modern astronomy is constructed independently of the idea of the ether. It is the physicists who, first through their studies upon light, then through the inductions drawn from them, have imposed upon science the idea of this universal fluid.* It may, accordingly, be asked if this idea will not be found in disagreement with the astronomical laws that have been established without it. Those who are averse to admitting the existence of the ether, do not fail to object that the progress of the stars must be retarded by this fluid; that the planets, by reason of the resistance they encounter, must constantly be approaching the sun, and that notwithstanding astronomers find no symptom of such an effect.

The retardation exists, perhaps, without it being possible to prove it, reply the partisans of the ether.

* "Whatever the eye perceives in the ether, the ear perceives in the air; whatever the ether presents to our organs by means of colors, the air presents to us by means of modulations and sounds. Thus Nature is always the same, always similar to herself, both in light and in sound, in the eye and in the ear; the only difference is, that in one she is quicker and more subtle, and in the other slower and more gross, exhibiting herself to our various senses by means of her various degrees and momenta, and being as perceptible to sense in one medium as she is in another. How admirable are the varied and sportive movements of nature! How charming and delightful does she render herself solely by her varieties in the motions of her elements, being as beautiful in the ether by the play of her colors as she is harmonious in the air by the modulations of her sounds! What gratifications does she afford to us in the diversified operations of her living machinery!" — *Swedenborg, Principia*, vol. ii. p. 309.

For a full and interesting account of the luminiferous ether, the reader is referred to the above-mentioned work.

If we enter the domain of facts, we are convinced that this retardation can only be very feeble, on account of the tenuity of the fluid which produces it. Calculations have been established, according to which the resistance of the ether would shorten, by three meters a year, the distance of the earth from the sun; the duration of the year would thus be shortened one second in six thousand years. The state of our astronomical observations does not allow the singling out of such a consequence from the midst of the perturbations of the terrestrial orbit already known.

Failing of any decisive facts in the planetary motions, the controversy is thrown back upon the comets. If the ethereal resistance is insensible for the planets by reason of their great density, it must be appreciable for the comets, which have, so to speak, no mass, and which have been termed *visible nothings.**

* It is not very long since the extreme tenuity of cometary matter has been established. Formerly the shock of these bodies was always considered dangerous for the planets. It is to a shock of this character that Buffon attributed the origin of our planetary system; a comet precipitating itself into the sun, had detached fragments of matter from it, and hurled them into space. Again, it is to shocks of this kind that various geologists attribute the terrestrial cataclysms; comets, coming in contact with the earth, would have displaced the axis of rotation and determined the great deluges. Such opinions no longer exist. Comets are regarded to-day as quite inoffensive heavenly bodies, in-

Here a consideration comes in to obscure the problem. The extreme lightness of the comets must render them sensible to the resistance of a universal medium undoubtedly; but it exposes them also to perturbations of other sorts. They are powerfully deviated from their course when they pass in the neighborhood of the planetary bodies. When Lexell's comet, in 1770, passed through the satellites of Jupiter, the time of its revolution became abruptly shortened from fifty years to five and a half years. How discern the influence of the ether in the midst of pertubations of this kind? Encke's comet, whose periodicity has been known since 1818, has a revolution of very short duration, about three and a quarter years; its orbit lies entirely within that of Jupiter. In comparing its successive appearances since 1818, there has been observed a gradual diminution in the time of its revolution. It has been proved, moreover, that this effect does not proceed from the perturbing influence

capable of disturbing the peace of the world. They have been seen to pass close by planets without causing any disorder in them. Twice has Lexell's comet been seen to rush through the satellites of Jupiter, without producing any derangement in them. According to the recent calculations of M. Faye, the nucleus of comets, which is the most compact portion, is scarcely nine times more dense than the air which remains in our pneumatic machines after we have made the vacuum as complete as possible; as to the density of the tail, it would be ten billion times less.

of the planets. Certain astronomers have concluded from this that it must be attributed to the resistance of a medium, and have there seen the first astronomical demonstration of the existence of the ether; but this conclusion, drawn from a solitary example, in the midst of the uncertainty still hanging over the greater part of the particulars of cometary motion, cannot be regarded as very binding.

Thus astronomical observations furnish no characteristic fact upon the subject of the resistance of a medium, and no conclusion in regard to it is to be drawn either from the course of the planets or that of the comets. But we have now to ask ourselves if the explanation just given upon the subject of the origin of attraction does not illuminate the problem with an entirely new light?

Mathematical analysis refers to two forces as the causes which produce the curvilinear motion of the stars. One, the initial force of impulsion or acquired velocity, tends to direct them in a straight line, while gravity incessantly deviates them from this course. This is that dynamic equilibrium, established by astronomers independently of any notion of the ether, which seemed to be compromised when the physicists admitted the existence of a universal medium; the ether must needs derange this balancing of two forces instituted without its aid.

If now we recognize the ether as the origin of at least one of the two forces, the question changes its aspect. It may no longer be said that it has remained foreign to the establishment of the equilibrium of the heavenly bodies; on the contrary, we find that we have unwittingly forced it to take part in this equilibrium. Henceforward let us no longer speak of a new resistance introduced by the ether! Its mode of resisting the celestial motions is precisely to determine the attraction, and so influence the course of the stars. We say that the ether produces gravity; that it urges the heavenly bodies in a certain direction; by so doing we have accounted for all the actions which it exercises, for the shocks it gives upon all sides. It would be a double task to introduce a second time into our calculations, under the form of resistance to motion, the shocks which the stars receive from the direction in which they move.

If this is so, if it is true to say that the ether cannot be considered at the same time as a cause of sidereal motion, and an obstacle to this motion, we need no longer be surprised that astronomy finds in no part of the heavens the mark of a resisting medium.

III.

Historical Notions regarding the Idea of Universal Attraction.

It is possible, then, to bring within the compass of our hypothesis the cause which produces the gravity of bodies; but this, we cannot disguise it, is one of the most difficult points we have to treat. Such is the power of habit over our minds, that the origin of attraction would seem to us unattainable. To connect this conception with a more general idea seems a chimerical undertaking. In order to support the demonstration we have attempted in regard to this, it will not be without use to give a slight sketch of the way in which this grand idea of universal attraction had its birth, and how it has been developed. By indicating the rôle which it has played in the history of our sciences, we shall better mark the place which it should hold in the physical science of our day. In beholding how the human mind has attained to a law so high, it will seem to us possible for it to go higher yet, and that gravity, in order to have explained so many things, cannot itself be inexplicable.

Modern astronomy begins with the book of the Revolutions of the Heavenly Bodies, which Copernicus published in 1543. Copernicus, overturning the doctrine of Ptolemy, placed the sun in the centre of the universe. Around this body he made revolve the six planets then known, Mercury, Venus, the Earth, Mars, Jupiter, and Saturn, and he endowed them also with a motion of rotation upon their axes.

Although dedicated to Pope Paul III., the book of the Revolutions of the Heavenly Bodies was condemned, as contrary to the text of the Scriptures.

Whether he desired to escape the censures of the Roman court, or whether he had the ambition to attach his name to a system which was peculiar to himself, Tycho-Brahe adopted an eclectic hypothesis. He deprived the earth of its double motion, and made the moon and the sun revolve around it, conformably to the doctrine of Ptolemy; but he admitted at the same time the revolution of Mercury, Venus, Mars, Jupiter, and Saturn around the sun. In spite of this whimsical theory, Tycho-Brahe is one of the founders of the science of the heavens. Assisted by his disciples and numerous colaborers in the little astronomical city which he had founded, he searched the heavens in all directions, and accumulated upon the subject of the planetary motions a prodigious quantity of observations, which served as a basis for the labors of Kepler.

The three great laws to which Kepler has given his name are well known.

Copernicus and Tycho-Brahe had preserved the faith of the ancients, who regarded the course of the planets as circular. It was upon this opinion that the attention of Kepler was first brought to bear. Comparing the observations of Tycho upon the motions of the planet Mars with those which he had himself made, he convinced himself that the orbit of this star was not circular; after having vainly tried several hypotheses, he finally discovered that he could satisfy the result of his calculations by supposing that the orbit of Mars was an ellipse, one focus of which was occupied by the sun. At the same time he found that the areas described around the focus by the radius vector are equal in equal times. Such are the two first laws pointed out by Kepler. After having verified them upon several planets, he published them in 1609, in a memoir entitled, *De motibus stellæ Martis*.

The third law is, that the squares of the times of the planetary revolutions are proportional to the cubes of the long axes of the orbits. It is this which cost the highest efforts of Kepler's persevering genius. The manner in which he announces this in his treatise, *Harmonices Mundi*, partakes of the enthusiasm which such a discovery caused him. "After having found," he says, "the true dimensions of the orbits through the observations of Brahe, by a long-continued and laborious effort, I have at length discovered the proportion of the periodic times to the extent of these orbits. And if you wish to know the exact date (of this discovery), it was the 8th of March, this very year, 1618, that, first conceived in my mind, then unskilfully attempted in figures, hence rejected as false, afterwards reproduced the 15th of May, with a new energy, it surmounted the darkness of my intelligence, and so fully confirmed was I in it by my labor of seventeen years upon the observations of Brahe, and from my own researches, that I first believed that I was dreaming. . . . But there is no longer any doubt; it is a very sure and very exact proposition, that the ratio between the periodic times of two planets is precisely sesquialter to the ratio of the mean distance." *

Thus Kepler had determined, in three truly great laws, the orbits of the planets and the conditions of their motion. He was so near the principle from which these laws are derived, that it may be asked if

* Half the long axis of a planetary orbit is often called the mean distance. It is, in fact, the mean between the greatest and least distance of the planet from the sun.

he did not foresee it. Endowed with an ardent imagination, he naturally sought the cause of these motions, the nature of which he had discovered; but in this relation his works show us hardly more than the exuberance of ancient astrological fancies. The old Pythagorean theories, the mysterious properties of numbers, here play a singular part; and one is astonished at the odd dreams which are mingled with the most serious calculations.*

He had, nevertheless, his theory upon solar attraction. He gave to the sun a movement of rotation upon an axis perpendicular to the ecliptic, thus foreseeing a truth which experience was only to prove somewhat later; immaterial forces emanating from this luminary in the plane of its equator, endowed with an activity decreasing in proportion to the distances, caused each planet to participate in this circular motion. The planet, carried along by this transcendent effluence, followed the rotation of the sun, and at the same time, by a sort of instinct or magnetism, it

* Yet it was *this mystical* part of Kepler's opinions, this belief in the mysterious properties of numbers, that led to a conviction, on his part, of a physical connection between the different parts of the universe, and finally to the discovery of those numerical and geometrical laws which govern them. — *Translator.*

alternately approached and receded from the central luminary, sometimes rising above the solar equator, and sometimes sinking below it.

At the same time that Kepler was determining the constitution of planetary motion, Galileo discovered the law of the acceleration of bodies which fall freely to the ground, or which glide over inclined planes ; he established the general properties of a uniformly accelerated motion.

The laws of gravity at the surface of the earth constitute the fundamental principles of mechanics. Ere long Huyghens perfected the theory of the pendulum, and gave, through his Theory of Central Forces in the Circle, brilliant suggestions concerning centrifugal force.

Such are the principal elements from which Newton derived the grand discovery of universal attraction. The methods of calculation had also just been enriched by some remarkable inventions. Descartes had founded the analytical geometry, and Fermat had just laid the principles of the infinitesimal calculus. Thus the labors of a half century, fruitful in great geometricians and great astronomers, concurred in bringing together the materials which Newton was able to employ.

Tradition relates that Newton, while in retirement

in the country, during the year 1666, saw an apple fall from a tree. Thereupon directing his thoughts to the system of the universe, he conceived the idea that the force which attracted bodies towards the surface of the earth was the one which made the moon turn around the earth, and the planets around the sun.

Kepler's laws furnished him with admirable data, from which he drew the consequences resulting from their analysis. From the law of the proportion between the areas and the times, he concluded that every planet is submitted to an attraction constantly directed towards the sun. From the elliptical motion, he concluded that for the same planet the tendency towards the sun varies from one point to another of the orbit in the inverse ratio of the squares of the distances. He had, then, the means of comparing the gravitation of any one planet towards the sun in any two points of its orbit; but this was not sufficient; it was necessary, besides, to know how to compare the gravitation of two different planets, for it might be that, passing from one planet to the other, there would be a change in the amount of attraction. The third law of Kepler, the proportion between the squares of the times and the cubes of the mean distances, permitted Newton to complete his theory, and to refer all these attractions to one. This law signified, in fact, that all the planets, of the same mass and at equal distances, would be equally attracted by the sun. The same equality of gravity exists in all the systems of satellites, and Newton assured himself of it in the case of the moon, as well as in that of the satellites of Jupiter.

It was with the lunar attraction that he began the verification of his theory. The question was to determine whether the force which incessantly deviates the moon towards the earth be identical with terrestrial gravity. In this case, the action of these forces referred to the centre of the earth would have to be in the ratio of the earth's radius, taken for unity, to the square of the distance separating the two heavenly bodies. Newton undertook this verification, starting with the experiments of Galileo upon heavy bodies; but there existed then only an inexact measurement of the earth's radius, and the great geometrician saw the result of his calculation in disagreement with his hypothesis. Thereupon, persuaded that unknown forces were joined to the moon's gravity, he gave up for a time his ideas. Some years later, the Academy of Sciences having just effected the measurement in France of a degree of the meridian, and a new measure of the earth's radius having resulted from this work, Newton recommenced his researches, and this time he found that the moon was retained in its orbit by the sole power of gravity. The sight of this result, of which he had despaired, caused him, so say his biographers, so lively an excitement that he could not verify his calculation, and was obliged to trust the care of it to a friend.

Thus one and the same law, a law unique and grand, explained all the motions of bodies on the surface of the planets, and those of the stars in space. The principal developments of this law were collected in the immortal treatise, the Mathematical Principles, which Newton published towards the close of the year 1687.

Having reached a principle which embraced the universe entire, Newton himself made brilliant applications of it. He proved that the earth in its rotation must become flattened at the poles, and he determined the amount of variation in the length of the degrees of the meridian. He saw that the attractions of the sun and moon give rise to, and maintain in the sea, those oscillations which constitute its ebb and flow. He demonstrated, finally, the mode in which the spheroidicity of the earth at the equator and the inclination of the polar axis to the ecliptic determine the phenomenon of the precession of the equinoxes. He recognized, in a general way, even with exactness upon some points, the perturbations which affect the planetary system. If a single planet be considered as gravitating towards the centre of the sun, it must obey strictly the laws of Kepler ; but this is no longer the case when it concerns the attraction of several of the heavenly bodies towards each other, if instead of two bodies there be three. The conditions are then changed and complicated, even so far as to become amenable to analysis only with great difficulty. Newton assigned the meaning, and sometimes the numerical value, of several of the planetary perturbations, thus tracing in their germ the methods which were in our day to make it possible for mathematical calculation to find the planet Neptune at the extremity of the solar system. He recognized those disturbing phenomena which affect the elements of the planetary orbits, and which astronomy divides into two categories, the *secular inequalities* at very long intervals, and the periodic inequalities, the period of which is only some years.

But when he saw that the planetary ellipses successively approached and receded from the circular form ; that their orbits did not always remain equally inclined to a fixed plane ; that they cut the ecliptic according to lines which changed their position in space, a discouraging thought entered his spirit. It seemed to him that the feeble values of all these variations, accumulating in the course of centuries, must overturn the system of the universe. He declared that this system did not possess the lasting elements of preservation, and that it would be necessary for a transcendent power to intervene from time to time to repair the disorder.

Leibnitz eagerly opposed such an opinion, and ridiculed this faith in an intermittent miracle. Newton retorted with railleries upon the subject of the doctrine of the pre-established harmony, which was, it must be confessed, one of the oddest conceptions of metaphysics. The quarrel soon became bitter, and was complicated with the sharp dispute in which were seen these two great minds contending for the invention of the differential calculus.

Newton had traced out a sublime draught of the theory of sidereal motion; but it was merely a draught. It was necessary for mathematical analysis to accomplish prodigies; it was necessary for Euler, Clairaut, D'Alembert, Lagrange, and Laplace to amass their efforts, that the sketch might become a completed design.

Clairaut first gave a complete and satisfactory solution of the problem of *the three bodies*, which consists in determining the course of a planet subjected to the combined attractions of two other heavenly bodies.

There continued to be an uneasiness in regard to the astronomical perturbations whose periodicity was unknown. It was Laplace who first discovered in them an evidence sufficient to reassure us as to the conservation of the planetary system. In the midst of perturbations of every kind which observation made known, there was a quantity which remained constant, or which at least was only subject to but slight periodical variations. This was the major axis of each orbit, upon which depends, according to the third law of Kepler, the time of revolution of each planet. The solar system became, as it were, confirmed, and it was seen that it only oscillated about a mean condition, from which it never departed save by very small quantities.

Scarcely had this result been obtained when it seemed to be compromised. Constant inequalities were pointed out in the journeys of Jupiter and Saturn. Comparing ancient observations with modern, it was found that the motion of Jupiter was constantly accelerating, and that of Saturn was subject to a gradual retardation. The theoretical consequence of these facts was of a character to strike attention. It was necessary to conclude from them that Jupiter would gradually approach the sun, and finally cast himself into it. Saturn, on the contrary, was destined to become farther and farther removed from the centre of our system, and to plunge forever into the darkness of the space which our telescopes do not reach. The Academy of Sciences was disturbed at these possible results ; it summoned to this question the labors of the geometricians. Euler and Lagrange descended into the arena, without solving the difficulty ; it was again the sagacious analysis of Laplace which demonstrated, in the reciprocal perturbations of Jupiter and Saturn, the reason of the anomalies pointed out by observers, and which explained them by an inequality of long period, the development of which demands more than nine hundred years.

There were observed, moreover, inequalities whose period was still longer ; those which depend upon the precession of the equinoxes have a duration of ten hundred and sixty centuries ; the eccentricity of the earth's orbit goes on diminishing from the most remote ages, obeying a period whose duration is reckoned neither by hundreds nor by thousands of years, — a duration of time, in which the history of astronomical observations, that even of the human race, figure but as a point.

We have now followed the Newtonian idea up to the time when it accounted for all the celestial phenomena ; but it must not be thought that this idea was accepted at once by all minds. Its beginnings were marked by the liveliest disputes. The quarrel between Cartesianism and Newtonianism filled all the first half of the eighteenth century.

The system of physics of Descartes yielded but slowly before that of Newton, and in the domain of facts itself the supremacy between the two doctrines remained for a long time undecided. Not the synthesis only, which Newton had drawn from Kepler's laws, but those very laws themselves were for a long time disputed. At the close of the seventeenth century, Dominique Cassini proposed to substitute for the ellipses of Kepler a curve, which seemed to be more accurately adapted to the sidereal motions ; this curve has taken the name of the *Cassinoid.** One of the first consequences of Newton's theory, the flattening of the earth towards the poles, was denied. The sons of Cassini, heirs of the paternal traditions, proved by the measurement of an arc of the meridian that the earth was a spheroid, elongated in the direction of its axis. This opinion prevailed in our Academy of Sciences up to the time when an expedition was organized for deter-

* In the ellipse, the sum of the radii drawn from a point of the curve to the two foci is constant. In the Cassinoid, a curve of the fourth degree, it is the product of the two radii, which is constant.

mining the comparative lengths of a degree near the pole and near the equator. Bouguer and La Condamine set out in 1735 for Peru; Maupertius and Clairaut betook themselves to Lapland. The hypothesis of Newton came out of this trial victoriously, and towards 1744 the greater number of savants, including the two Cassini themselves, recognized the errors in the experiment or reasoning which had made them take the earth for an elongated spheroid. Then it was that Descartes' system of physics was finally overturned, together with a great part of his metaphysics, and the Newtonian idea, popularized by Voltaire, and afterwards by the Encyclopedists, remained triumphant.

But, if we are going to the bottom of things, what was the point especially discussed between the Cartesians and the disciples of Newton?

Descartes set out with this principle, that the whole universe is to be explained by motion, and we, at least, shall not make this a reproach against him; it is upon this very stand-point that contemporary science places herself, and the majority of our physicists, whether they will or no, are found to be Cartesians in this respect. But Descartes, his principle laid down, without facts, without observations, without experiments, without proof of any sort, by a pure conception of his mind, had created a system of the universe; the universe was full of matter, absolutely full, without any void; vast circular currents existed throughout this matter, and carried along with them the planets, as the current of a river carries vessels along. The disciples of Descartes abated nothing from the idea of their master, and they heaped up laborious explanations in order to show how the vortices could be propagated in a space absolutely full; how the particles of matter could glide over each other without any interstitial vacuum.

To confront this doctrine, Newton brought forward the law of universal gravitation; this law contained in itself an enormous mass of facts. Not only did it explain all that were already known, but it foreshadowed new ones, and experiment justified these foreshadowings. The Newtonians then felt themselves upon very solid ground. In their enthusiasm they left the domain of facts, and came to look upon gravity as a mysterious cause, of an order superior to all physical phenomena. Newton, as we have just now shown, guarded himself from this excess, at least in the beginning of his career, and in the book of Principles. Perhaps he was less reserved in this respect in his old age. As to his followers, they had an evident tendency to believe themselves in possession of a supernatural principle.

It was against this tendency that the Cartesian school reacted; it rejected the hidden cause presented to it; but it rejected, at the same time, both the cause and the effects demonstrated by experiment. It closed its eyes in order not to behold the new astronomical system, and it obstinately persisted in its fanciful system of physics. It fell into ridicule, and the hypothesis of vortices, of which Fontenello was the last defender, drew down in its fall the whole Cartesian doctrine.

Thus one often sees, in the conflict of human ideas, when two great doctrines are violently opposed to each other, one of them succumb entirely, and the conquerors effacing, without distinction, all that the conquered had inscribed upon their banner.

As for us, who now regard this historical debate through the softening influence of years, we see the ground upon which the two hostile doctrines might be reconciled. The gravitation of Newton, and all the facts which it embraces, appear to us in harmony with the Cartesian principle, as consequences of the motions of matter.

IV.

Hypotheses in regard to the Formation of Worlds and the Origin of Gravity.

The Newtonian principle and the Cartesian principle, so long hostile, became united, and merged into each other, in the general idea we are now able to form concerning the system of the universe.

This general idea is formed in our minds when we comprise in one general view the hypothesis of Laplace, concerning the birth of the solar system; the conjectures drawn by contemporaneous astronomy, from the appearance of the nebulæ, and the facts we have developed above with regard to the function of an ethereal substance.

Let us begin by recounting, in a few words, the hypothesis of Laplace.

Our planetary system was at first only a nebula; its limits extended far beyond the present orbits of our planets, and it became successively condensed in the course of ages.

Laplace sketches, in grand outline, the history of this gradual condensation. A solar nucleus first forms in the nebula. This nascent sun is a gaseous mass, endowed with a motion of rotation, which it shares with an immense atmosphere. By the general cooling of the system this atmosphere leaves successively, in the plane of its equator, lenticular zones, from which spring the planets. Sometimes these zones preserve the circular form, of which the rings of Saturn afford us examples. Most frequently they separate into several parts. The fragments may remain ununited, as we observe in the system of minor planets, situated between Mars and Jupiter. They may also — and this is oftenest the case — be united together into a single mass.

The planets thus formed, are originally gaseous masses, which continue to turn about the sun; they turn also upon themselves, because, in the original ring, the molecules farthest removed from the solar centre had a greater velocity than the rest. By this rotation each of them takes the form of a spheroid, flattened at the poles, and very soon, in these miniature worlds, is begun anew the phenomenon just now explained. The planetary atmosphere gives off rings, from which spring the satellites.

The nuclei of the planets and those of the satellites become hard at the surface, the atmospheres become denser between the nuclei, and the immense expanse, which was first filled by the nebula, is no longer occupied, save by a few celestial globes, which move regularly around their common centre.

The author of the *Mechanique Celeste* has only put forth this pretentious hypothesis with reserve; he has modestly placed it in a note at the end of his Exposition of the System of the Universe. It has not failed to assume a high importance, for it is the only conception which accounts for the principal planetary phenomena. It explains why all the planets circulate about the sun almost in the same plane; why this plane of general circulation is exactly that of the solar equator; why the planets describe ellipses which nearly resemble circles; why their motions of progression and of rotation take place in the same direction; why all the circumstances observed in the journey of the planets around the sun are reproduced in the circulation of the satellites around their planets.*

* Is there need of recalling here the brilliant experiment to which a Belgian physicist, M. Plateau, has attached his name, and which reproduces the principal phases of these creations of the heavenly bodies? In a vase there is placed a mixture of water and of alcohol, in the centre of which is put a drop of oil. Into this drop is introduced a needle, to which a regular motion of rotation is imparted. The oily sphere turns on its axis, and becomes flattened at the poles. Soon, from the expanded equational regions, if the experiment is well conducted, there escapes a sort of ring, which breaks into globules, each of which begins to turn about the central mass. One may thus construct a uni-

The hypothesis of Laplace leads us, then, from the origin of the sun to the complete development of our solar system; but let us now conceive of a phase anterior to this, and attempt to portray its history.

Let us go back to a period, in the succession of ages, when no system as yet existed.

The ether alone filled all space with its atoms in motion.

If this medium is strictly homogeneous in all its parts, a uniform vibration will continue without end; but if, among these primitive atoms, there exists, at certain points, some dissimilarity, the preponderating atoms immediately become centres of aggregation. They approach each other in the manner we have described. A sort of selection is thus effected throughout the universal mass; the ether becomes more and more homogeneous in proportion as the dissimilar elements become united at certain centres. Thus, there is formed in the midst of the ether, now become more

verse in a glass of water. The reader will observe, on a little reflection, that this experiment is not a fair illustration of Laplace's theory. The oily ring is thrown off by the *centrifugal* force of the revolving globule, while in the nebular hypothesis the rings, out of which the planets are formed, are successively abandoned by the cooling and *contraction* of the nebular mass. — *Translator.*

and more purified, a cosmical essence universally extended, the subtle germ of ponderable matter.

It is, in fact, gravity that has just taken its origin in the phenomenon we have sketched, and it becomes more clearly pronounced in proportion as the molecular groups are better defined, and the ether reduced to a state of atomic uniformity.

Here, then, is space occupied by a sort of embryonic network, the interstices of which are filled by the ethereal atoms. The motion of attraction thus begun never ceases.

At the same time that the ether was tending towards a uniform condition, the rudimentary molecules were absorbing all the unlike elements; thus they have been unequally pressed upon in different directions; the motions of progression and rotation follow as a natural consequence.

Variety is also characteristic of the cosmical network, from the very nature of its origin; it becomes torn then into irregular shreds here and there, where the effects of concentration are manifested.

We here reach a point where telescopic observation comes to the aid of pure speculation. The farther artronomers penetrate into the depth of the heavens, the greater is the number of these cosmical bodies which they discover, some of which are resolvable into stars, while others preserve the appearance of irreducible nebulosities. Do these latter owe this appearance to distance only? and are we to believe that by the aid of stronger lenses they could be decomposed into luminous points? Opinion may vary in regard to this in such or such particular case, with reference to such or such especial nebula; but the aggregate of observations leads to the belief that these agglomerations are worlds in various stages of formation. In some, the cosmical matter would be still diffused; in others, the solar nuclei would be more or less formed; in others, again, the suns would have already generated their corteges of satellites.

Thus we should have before us, more or less accessible to our telescopes, specimens of the various phases through which worlds pass.

Greater importance will not attach to these suggestions than they deserve. If Laplace deferred his hypothesis to the end of one of his books, where shall we assign the place of the cosmical outline just sketched? We have attempted to carry back to the very origin of cosmical formations that conception which represents to us gravity as a consequence of the motions of ether. The views we have given in this connection may seem unjustifiable. They may be rejected without at the same time weakening the considerations which bear upon the nature itself of gravity, such as we can observe it to be in our own world, in the midst of circumstances accessible to our analysis.

V.

The Molecular Agencies, Cohesion, and Chemical Affinity.

WE must now return, by an abrupt transition, from the motions of heavenly bodies to the molecular phenomena; from the immense spaces to which gravity extends to the infinitely small distances in which cohesion and chemical affinity display themselves. We have already pointed out the enormous power of these last two forces; but the numerical results we have mentioned give only a faint idea of it. It is known that changes in cohesion, the freezing of water, for example, and the solidifying of bismuth, may shatter iron bottles of the thickness of several centimeters; we say nothing of the formidable effects produced by the action of affinities in explosible mixtures; the simple reactions which form and maintain the ordinary aggregations of matter have a power so great that they have been called, in figurative language, giants in disguise.

It seems at first thought that the heavenly bodies, in their journey through space, must use up the largest part of the *vis viva*, or living force, diffused throughout the universe; the contrary is true. The living force, represented by the motions of the heavenly bodies, is very weak compared with that concentrated in the molecular activities.

Before making a new step in the examination of these activities, it is important that we recur to the idea itself of the molecule, and that we define its meaning. The molecules of bodies reputed to be simple, such as oxygen, hydrogen, carbon, are they indivisible unities, veritable atoms, or are they aggregations?

This latter hypothesis, we have already said, alone seems admissible.

After the first labors which established the science of chemistry, when analysis stopped before a certain number of substances which it could not decompose, one was led to look upon these substances as different in their very nature. Such was the doctrine of Bergelius. By this theory carbon, gold, and platinum are bodies entirely heterogeneous, their atoms enjoying special and peculiar properties. Nevertheless, the notion of *equivalents*, introduced into chemistry at the beginning of this century, naturally inclined men's minds towards a different doctrine. When it was seen that simple bodies combine and replace each other in their combinations in clearly defined proportions, the equivalent quantities of different bodies came necessarily to be regarded as different collections formed out of one and the same substance.

Prout was the first to give definite shape to this opinion. According to him, the equivalent weights of simple bodies were the multiples of that of hydrogen. It was very soon observed that this law could not be maintained in the exact terms of its annunciation. The precise determination of certain equivalents was not in agreement with it. The first exceptions were made to disappear by taking as unity the half equivalent of hydrogen; but new difficulties arose, and it was necessary to have recourse to a more complicated division. Prout's law has thus lost, little by little, its primitive originality. It remains, notwithstanding, corrected by necessary restrictions, as an important agreement in favor of the elementary unity of bodies.

We have already shown how the new physics goes back to the very ethereal atoms for this elementary unity. Between the molecules of oxygen, of hydrogen, of carbon, of gold, of platinum, it conceives of no difference which bears upon the quality of matter; it beholds only in these different bodies the properties which result from motion. If this is true of these bodies compared with each other, it is true also of the same bodies compared with the ether. Between them and the ether where could there be found a difference which should have an influence upon their material quality? Thus every elementary molecule appears to us as formed of ethereal atoms. Heat disorganizes bodies; it goes so far as to separate the hydrogen and the oxygen, which form the vapor of water; there would be a final step to make; by additionally heating these same molecules, they might, doubtless, in the end be driven apart, and be resolved into ethereal atoms, either directly or by successive steps.

Here, then, is the manner in which the scale of material aggregation presents itself to our eyes. In the state of extreme tenuity, the ethereal atom; then comes the elementary molecule of bodies regarded as simple; these molecules combine, and from them result compound or chemical molecules. These last unite in their turn, and thus form the particles of bodies.

It may be conceived, at least in a general way, how the aggregation of an elementary molecule may result from the activity of a medium and the relative motions of its parts. Without insisting upon this point, we can represent this order of phenomena by means of several rough examples, a few far-fetched analogies. It is in this manner that the pressure of the air maintains, in opposition to each other, the segments of a hollow sphere. It is thus that a liquid jet often assumes the appearance and the consistency of a solid through the common motion of its parts. It is thus that we often see eddies of wind or of dust pass over long distances without losing their shape, because the elements that compose them are endowed with the same angular velocity.

It is equally true, if one examines these questions closely, that the new physics, in the light it throws upon them, only reveals to us a few fugitive outlines. We should demand of it in vain to show by some decisive examples how the various properties of molecules arise from a combination of motions. This diversity, which springs, so to speak, from the very bosom of matter, always has been, and still remains one of the strangest phenomena which man can investigate.

Early science saw in bodies a kind of duality; it imagined, on one part, a matter deprived of qualities of its own, but capable of receiving them all, and on the other essences, which joined themselves to bodies in order to constitute their properties; it supposed that these essences could be isolated by distillation, and the alchemist strove to collect them, in order to infuse them into matter at his will.

After the doctrine of essences, there prevailed the idea of forms; an æsthetic principle, concealed in the interior of bodies, determined the moulds in which the molecular diversity was produced. Let us remark that this conception approaches that of motion. The idea of motion does not ultimate itself without a certain idea of form; geometry determines the curves, and the ideal surfaces in which the motions are produced, and by which they are limited.

The new physics refers to motion the structure and properties of molecules. It draws this conclusion from the aggregate of laws which it has discovered; it believes itself authorized in this by what it knows of several great natural phenomena; by what it has learned of light, heat, and electricity; by the inductions to which it has been brought regarding the nature of universal attraction. But the future alone will show whether it can reach the original conditions which diversify the motions taking place in the hidden depths of matter.

We must not occupy ourselves longer with the metaphysics of molecules. The principal results we have successively enunciated are independent of every hypothesis concerning the constitution of molecules.

We have been careful to reserve the term Atom for the elements of the ether, and to apply the name of molecule to those of ponderable matter; but otherwise, throughout the course of this work, if certain incidental theories, which are easily separated from it, be put aside, we may preserve the primitive notion furnished by chemistry, and regard the elementary molecules as little indivisible blocks, whose interior construction possesses no influence over phenomena.

It has been seen how molecules, immersed in ether, come to attract each other. We require a new principle to explain cohesion, and we find it in the hypothesis of molecular rotation, from which Father Secchi has drawn so many ingenious results.

In their rotation molecules must carry along with them an atmosphere of ethereal atoms; this is a fact we have already made prominent when treating of the change of state of bodies. The existence of these atmospheres — and here we must avoid a possible confusion — is entirely distinct from the phenomenon which distributes the ether in layers of different density. This latter effect extends to infinity; the former takes place only in a very limited space, in the immediate vicinity of the molecule. In this space the atoms share directly in the molecular rotation; outside of it they are independent of it. It has been shown above how these atmospheres behave when a body, losing its heat, is brought from the gaseous state to the liquid form, and from this to the solid condition. Let us remark here, that this hypothesis explains why the liquid and solid states take place all at once, at a given moment, when the molecules have been brought within a fixed distance of each other. So long as the atmospheres do not touch, no trace of cohesion shows itself; when they meet, this force arises. We understand also why the temperatures of melting and solidifying are fixed for the same body; these effects take place at the precise moment when the atmospheres, varying with the temperature, have attained the desired diameter.

Now what is affinity? Let us note the nature of its action. It acts for a time, more or less precipitately, in order to disturb an equilibrium; the bodies concerned saturate each other; then a new equilibrium succeeds. This phenomenon may be explained by the very hypothesis we have just made use of.

Between homogeneous molecules all the atmospheres are alike, and there is no cause determining one to modify the other. In this case cohesion is produced. If, on the other hand, molecules of different kinds confront each other, there is a variety in the atmospheres; these may penetrate each other, and modify in this way the position of their respective molecules.

In this way the principle of chemical affinity becomes known to us.

The more unequal are the atmospheres, the more opportunity will there be for the equilibrium to be destroyed, and the greater will be the energy of the chemical action. They may differ, besides, not only in their volume, but also in their velocities, and they thus present several elements of variation. Temperature naturally influences the state of the atmospheres, and thereby changes the conditions of affinity. It may happen that two molecules, which have at a certain moment dissimilar atmospheres, and consequently a very great affinity, shall come to have, if the temperature changes, like atmospheres, and consequently a moderate affinity. It may even result, if the temperature continues to vary, that the relative value of the atmospheres will be reversed. Known anomalies could thus be accounted for; in this manner could be explained why, at temperatures very near each other, we sometimes see iron decomposing water, and setting the hydrogen at liberty; and sometimes, on the contrary, the hydrogen decomposing the oxide of iron in order to get possession of the oxygen.

The chemical molecule has then a general envelope, but that is not to say that the elementary molecules remain without their own minor atmospheres. It must be observed, moreover, that these general atmospheres are, so to speak, external phenomena, in which we naturally find again the motions themselves of the molecules. Between all these motions, those of the molecules, those of the partial envelope, and those of the general envelope, an equilibrium is established, from which results the stability of the combination. The compound will be so much the more stable as this dynamic equilibrium shall have less chance of being disturbed. If the elements are numerous, a slight variation of temperature puts disorder into this aggregation, and destroys its bonds of union.

This effect shows itself more clearly in proportion as we go from the mineral kingdom, where a certain simplicity reigns, to organic substances, the structure of which is more complicated. It is said that a molecule of albumen contains nine hundred elementary molecules. It is conceivable that compounds so complex may be easily destroyed by variations of temperature. The complication is much greater yet in organized tissues. Thus vegetables are confined each to a particular climate, and if animals are able to live in regions more extended, it is because they carry in themselves a source of heat which renders the temperature of each almost constant.

From the time of Lavoisier, the science of chemistry has been constructed upon the idea of masses; its relation to velocities remains wholly to be established. Now masses and velocities form two series of elements, which it is equally necessary to be acquainted with in order to appreciate the living forces with which molecules are endowed, and the varying results which they may thus give rise to.

We cannot refrain from remarking, however, that chemistry has made universal progress through the consideration of masses only. The law of *definite proportions*, the law of *multiple proportions*, even the notion of chemical equivalence, which naturally resulted from these two fundamental laws, are independent of all idea of motion. By means of their relative weights simple bodies have been followed into their elementary combinations, and the scale of their saturation determined. Later, organic chemistry becomes founded, through the study primarily of fatty substances, afterwards through the first analyses of the alcohols and ethers; the scales then prove insufficient for pursuing the complication of phenomena, and yet the theories which chemistry gives rise to seem, at first sight, to be applicable only to molecules in a state of repose. The law of *substitutions* sums up the progress of organic chemistry. This law does not imply, so at least one might think, any idea of molecular motion; it may be understood as the mutual replacement of partial groups among motionless groups of atoms, if one is satisfied with a summary view of it. But we no longer need show what incompleteness such a manner of appreciating chemical phenomena would possess. It would not be possible to compare molecular structure to the superposition of the stones of a building; if it is to be represented by an appropriate figure, we must liken it, on a limited scale, to solar vortices which should penetrate each other, and whose elements should take on in the encounter a new and unstable equilibrium.

Furthermore, we are not here dealing with a merely theoretical conception. If we return to the domain of facts, we see that chemical action produces a work, a result peculiar to masses, endowed with velocity. It is chemical action; it is the combustion of coal that sets the most of our engines in motion.

Just now we do not know how to make a direct measurement of chemical work; we determine it only through the medium of heat or of electricity; by these indirect means we already obtain an appreciation of it sufficiently exact. We judge of chemical action by its exterior effects, and it is a result not to be despised. To know it in itself, to penetrate its secret, to comprehend its interior play, we must determine the velocities as well as the molecular masses. If we were in possession of the terms of this twofold quality, we should see disappear all that chemistry yet retains that is whimsical and capricious; we might explain the different combinations, and the material properties resulting from them.

Then would be established molecular mechanics, which would comprehend in their entirety, not only chemical phenomena, but all the natural phenomena we have successively treated of, those of gravity as well as those of heat, those of electricity as well as those of light; a universal system of dynamics would embrace astronomy, physics, and chemistry.

CHAPTER VI.

LIVING BEINGS.

I.

Vital Activity consists in the Transformation, not in the Creation of Motions.

We have nearly exhausted the programme we had marked out in advance; we have directed our attention successively to the principal phenomena pertaining to the physical sciences; we have shown their mutual relations, and pointed out their fundamental unity. We could here terminate our examination, and consider our task as accomplished; the results we have reached now appear in their general bearing. We have not yet, however, paid any attention to living beings, which also form a part of the physical universe. Should they also be comprehended in the phenomenal unity upon which we have fixed our attention, or should they be excluded from it? Do they wholly obey the laws, whose mutual connection has been shown, or, if they are independent of them in some respects, what are their immunities?

The simple enunciation of these questions calls to mind the vast problems that have, from time to time, agitated mankind; so many theories concerning the nature of life, so much effort laid out on human personality, so many discussions in regard to the principles of a higher essence! Let it not be expected that we shall attack these lofty questions. We may reserve them intact, and we need not venture upon the field of transcendental speculation in order to show how the great law under which we have brought the operations of nature is verified also in organized beings.

It seems, according to the labors of modern physiology, that in the cell must be sought the principle of vital activity. Vegetables, like animals, are composed of cells.

Every vegetable is composed of an association of little sacs or vesicles which assume, in crowding together, the polyhedral form. Each one forms a closed organ which has its own life, and which is, as it were, the integral part of the vegetable.

The case is not otherwise with animals; but the more perfect the general organism, the greater the variety observable in the cells. In the lower degrees of the animal scale, among the infusoria of the lowest species, are found creatures of the simplest composition that it is possible to imagine. The cells, all of them, entirely similar to each other, fill an envelope furnished with vibratile cilia, by the aid of which the animal moves. With the higher animals, with vertebrates, with man, there are great differences between the cells belonging to the tissues of different organs. A nucleus, more or less complex, at the centre, a fine membrane at the periphery, between the two a liquid, either simple or compound, — such are the constituent principles of the cell, and they exhibit a sufficient variety of elements to display very great dissimilarities between the cells composing the different muscular fibres, the various nervous filaments, the mucous and serous membranes, &c. In the midst of this diversity, each cell possesses, in the interior of the collective being, a relative independence, a sort of autonomy. Every family of these vesicles has its own government, its food, poisons, diseases.

Moreover it is known, since the ingenious discoveries of Dutrochet, how these little sacs are nourished, though entirely closed, separated even from each other by the double partition which results from their back to back arrangement; it is known that they succeed in absorbing the liquids outside of them, and in expelling, in part, those with which they are filled. This phenomenon of endo-exosmosis, together with capillary attraction, is sufficient to account for the ascent and descent of the sap in vegetables. It shows how in animals the different cells may incessantly renew their contents, and obtain, by an elective straining process, all that is needed for their support. Not only are these vesicles nourished, thanks to this mechanism, but, by an action communicated from one to the other, they succeed in drawing up liquids through canals that have no opening, and in pouring them into other canals equally closed, thus establishing throughout the mass of the tissues a capillary circulation, the principle of which for a long time escaped all research.

Thus we find, at the origin of life, cells which form the primary bases of organization.* It may be said that they constitute, in the two organic kingdoms, individuals that may be compared with the atoms of the mineral kingdom; atom, individual, two terms bor-

* Concerning the character of the life of these ultimate cells, see Physical Basis of Life, T. H. Huxley. The substance, *protoplasm*, which is considered in this treatise as forming the structural unit of all vegetable and animal life, contains only the four elements, carbon, oxygen, hydrogen, and nitrogen. — *Translator.*

rowed from different languages to express the same idea.

But do we know how the cells are produced? and has the secret of their formation been discovered? We have seen in the germ of vegetables a primary cell nourished by the starch contained in the grain, and converted by the process of germination into dextrine and sugar; we have seen new cells arrange themselves beside the first one by a process of *germination* or budding; the soluble substances, elaborated in this rudimentary life, thus come to constitute the primary elements of vegetables. In the animal germ, in the egg, we see a granular matter dividing itself into several spheroidal segments, and each of these converted into a vesicle by the coagulation of its superficial layer; then the vesicles attach themselves to each other, multiply themselves by division by the formation of interior membranes, and finally constitute the cellular tissue, out of which is to come the embryo. It is in this tissue that are arranged, by an analogous mechanism, the rudiments of organs, of a circulatory apparatus, and a nervous system.

As to the elements themselves which compose living organisms, vegetable or animal, it is here unnecessary to recall the fact that they are all borrowed in their last analysis from the inorganic world. As we ascend the animal scale, we find an increasing complication in this respect, notwithstanding there never enters into living beings more than a very limited number of simple bodies. The human body, the most complex of all, comprises fourteen simple bodies — oxygen, hydrogen, azote, carbon, sulphur, phosphorus, fluorine, chlorine, sodium, potassium, calcium, magnesium, silicon, and iron. However complicated may be the architecture of the molecules, the entire man is reducible to these fourteen elements.

If we now attempt to condense the primordial notion of life, such as it results from these suggestions, if we strive to reduce it to its essential principles, what do we find?

On the one side, the materials themselves of the inorganic world.

On the other side, a series of motions, which succeed each other in a determined order.

The definite succession of these motions doubtless exhibits a character entirely unique, but throughout their successive transformations there will be found nothing in them which jars with the laws of molecular mechanics.

Do we mean that we have here all the elements of life? What is the cause which forms the first cellule, which produces from it the development of the being, which regulates and limits its evolution? In view of the facts, it is too evident that we cannot reply to this question. We have, then, only two courses to pursue — either to suspend our judgment, or to admit a special cause, the principle of which is peculiar to vital phenomena.

With the nature itself of this course we have not to concern ourselves here; and, since it is manifested by means of motions, its name is to be found in the language we speak, we must call it a force.

What do the preliminary considerations we have just laid down teach us concerning the action of this force? Upon this point there must be a thorough understanding. It determines motions, but it can only produce them at the expense of anterior motions; just as it does not create the materials of organisms, but merely shapes them by the aid of pre-existing elements, so it does not create motions, but only transforms them. It is thus that vital phenomena, without losing their special character, enter entirely into the category of material motions. If the vital force has a peculiar activity, this activity consists in transforming, not in creating. It furnishes us, then, with a fresh confirmation of the great law whose development we are seeking throughout the entire universe.

Such is the stand-point to which we shall constantly be brought whenever we consider the phenomena of life.*

* The tendency of modern science is to include in the same category all displays of force. whether exhibited in organized beings or inorganic things. Heat, light, chemical affinity and *vitality* are thus mutually correlated. The idea of a distinct principle of vitality is, however, openly claimed by many, and tacitly acknowledged by most scientific writers of the present day. But the old boundary lines between the animal and vegetable kingdoms it has been necessary to abandon, since the microscope has disclosed by its keen scrutiny that no such lines exist. Even the power of voluntary motion, supposed to be the exclusive privilege of animals, has its analogies, if not its exact counterpart, in the spontaneous movements of many species of plants. The distinction between the mineral kingdom and the other two, between the inanimate on the one hand and the animate on the other, is equally ill-founded, for the truth seems to be that all created things possess life, each in the degree and kind corresponding to its use and destiny in the order of creation. Likewise the subdivisions of these kingdoms, with their systems of classification, prove but temporary make-shifts, which need to be constantly modified. The real fundamental distinctions which undoubtedly exist between the three kingdoms of nature, appear unattainable by science, whose province is to deal with effects, and not with causes. Our author has thought best not to touch upon these vexed questions; but we must admit that his assertion that vital *phenomena*, as being but manifestations in matter of an interior active principle, can be included in the category of material motions, is fully sustained by the results of modern investigation. — *Translator.*

II.

The Manner in which the Laws of Thermo-dynamics are Verified in the Case of Animated Beings.

THE respiration of animals, the circulation of the blood, nutrition, all contribute to a production of heat. A production of heat is the grand result of all these functions. Now direct observation has succeeded in tracing this into its essential conditions, and in showing how the heat is produced in accordance with the principles of Thermo-dynamics.

First, let us regard the state of repose; let us consider the condition of a man who performs no external work.

Animal heat results from the slow oxidations which take place within the organism. It may be added that it is due almost entirely to the combinations of oxygen with hydrogen and carbon. It is easy, then, by comparing the gases which enter and come from the lungs, to calculate the number of heat-units which a man produces in an hour. This is found to be an average of one hundred and twenty heat-units, with a variation, according to the subject, of about a third of this total amount.

What becomes of the heat-units thus produced?

The man must necessarily lose them as fast as developed, since the temperature of his body remains constant.* He gives it forth, in fact, under several forms, — pulmonary and cutaneous evaporation, heating of the expired air, radiation, contact with external objects. If the amount of heat emitted from the body in these various ways be directly measured, it will be found to equal that which is produced within the body; and so observation confirms the preconceptions of the theory.

Let us observe that, in the loss to be established between man and the surrounding medium, we have not reckoned the work which is accomplished in the interior of the body. The heart, for example, operates constantly after the manner of a force-pump; it acts incessantly with a force that may be estimated at the seventy-fifth part of a single horse-power, and its action thus represents the effect of nine heat-units per

* This temperature, as is known, is about thirty-seven degrees Centigrade. Climates exert no influence in this respect; between the inhabitants of the warmest countries and those of the coldest there is found hardly the difference of a degree. The kind of food has itself no action upon human temperature. In India it is likewise thirty-seven degrees with the native workmen, who eat only rice and fish; with the priests of Buddha, who live on vegetables, and the soldiers, fed chiefly on meats. A variation of four or five degrees in the average temperature of the human body constitutes a pathological condition which speedily terminates in death.

hour. Many other interior motions take place that might be estimated in the same way with more or less exactness; but the cycle of these phenomena being accomplished entirely within the body, there is an interior equivalence between the quantities of heat and the work they represent, and they are not taken into account in the exchange which takes place between man and the surrounding medium.

So much for the state of repose. Let us now consider the man who executes movements, and who produces an external work.

The beautiful researches of M. Hirn have shown that in the human body heat is transformed into work, and work into heat, according to the numerical ratio which we have already so often introduced; a unit of heat is converted into four hundred and twenty-five kilogrammetres, and reciprocally.

M. Hirn has selected for the object of his investigations the work which a man produces in raising his own body. When we ascend a slope, or when we descend it, our muscular force and gravity are put in antagonism. Practically this antagonism is complicated with the horizontal reactions due to the friction necessary for walking. M. Hirn, by an ingenious contrivance, has succeeded in eliminating this source of complication, leaving only the vertical forces to be taken into account. Imagine a man moving along the steps of a movable wheel; if the wheel be suitably turned, the man, without really having to change his place, will realize the artificial conditions of ascending, descending, and level walking, in which vertical actions alone are put into operation. The subject of his experiment would produce an external work when displacing the centre of gravity of his body in order to reach a higher step; if he descended, on the contrary, his weight would operate as if he had received external work, and his body would profit, in some sort, by a certain amount of motor force; if he walked without ascending or descending, his centre of gravity would be alternately raised and lowered in equal degrees; there would be an equivalent production and consumption of external work.

The theory clearly indicated the calorific effects that should be exhibited under these different circumstances, and they were produced in such a way as to fully justify the inductions of the experimenter. M. Hirn had first established, by direct measurements, that in a state of repose every gramme of oxygen absorbed invariably disengaged five heat-units; next, observing the state of motion, he saw that this proportion varied. If a man, weighing seventy-five kilogrammes, raised his weight four hundred and twenty-five metres, each gramme of oxygen disengaged less heat, and seventy-five heat-units, the exact representation of the work produced, disappeared. If the same man descended four hundred and twenty-five metres, each gramme of oxygen disengaged more than five heat-units, and the descent thus left in the organism seventy-five units, which could not be attributed to respiratory action. Moreover, respiration continued to yield five heat-units per gramme of oxygen in the case of level walking.

These striking results have been confirmed by a series of repeated experiments.*

At first thought, one may be astonished that walking on a level, as regards work, leads to no expenditure, and that descent constitutes, in this respect, a sort of gain, seeing that both of them — even under the conditions employed by M. Hirn — demand certain efforts, and result in a certain amount of fatigue. More than this, the case of ascent may even give rise to an apparent objection. How does it happen, it may

* It has been estimated that a man of average weight produces, in the climate of France, three thousand two hundred and fifty heat-units per day, or a sufficient amount of heat to raise seven gallons of water to the boiling point. (See an article from the French of Fernand Papillon, in No. 10 of the Popular Science Monthly, for interesting facts in regard to animal heat.) — *Translator.*

be said, that the act of ascending consumes heat when the body is manifestly warmed in producing this work? It is important to do away with these apparent contradictions, which would be of a character to leave in the mind a vague distrust of the theory just unfolded.

Yes; the work corresponding to the act of ascending consumes heat, but at the same time it accelerates the respiratory action and circulation; the volume of inspired air increases, and the absorbing power of the lungs is raised in a proportion often considerable. The quantity of oxygen absorbed, consequently the heat produced, increases even to five fold. M. Hirn has established these facts by putting himself into the apparatus which he employed for his experiments.

For an ascent of four hundred and fifty metres per hour, the number of pulsations of the heart was raised from eighty to one hundred and forty; the number of respirations a minute went from eighteen to thirty; the body of respired air in an hour was augmented from seven hundred litres to twenty-three hundred. As a result of this increasing activity in the respiration and circulation, the experimenter consumed no longer thirty grammes, as in a state of rest, but even one hundred and thirty-two grammes of oxygen per hour. Thus, in spite of the consumption produced by the work, an excess of heat is developed within the body, and the individual becomes warm.

Considerations of a like character would do away with the difficulty which we pointed out with regard to level walking and descending. To speak only of the first case, every step is divided into two periods; in the one, the weight of the body is raised, in the other it is lowered. The first period consumes heat; the second restores an equal quantity. Regarded in this light, the calorific equilibrium is not disturbed; but the organism, responding to the call of the muscles, alternately contracting and elongating, develops an excess of heat. This excess may be sufficient for an interior work of the muscles from which fatigue may arise, but which, according to an example already given above, is not at all to be considered in the exchange effected between man and the surrounding medium.

The mechanical theory of heat is then confirmed and illustrated in the human motor, as in all others. The man who, in M. Hirn's experiments, gave the best dynamic results restored in work in the ratio of twelve to one hundred of the heat produced; this is nearly the amount yielded by our most perfect engines.

If we follow up this paralellism between the weight

of the motor and the power it develops, we still find a sort of equality between man and our engines ; but animated nature presents us, in this respect, a class of beings especially favored. These are the birds.

These wonderful motors evolve a force of one horse (steam), while possessing a weight of only five or six kilogrammes. Their physiological structure, together with their relative lightness, gives them the means of enduring the enormous work they are obliged to produce in order to sustain themselves in the atmosphere. The bird is a centre of combustion of exceeding activity ; his whole body is, so to speak, but a lung ; the air, powerfully solicited through the very play of the wings, enters abundantly to vivify the blood, which the heart impels with prodigious power through the vessels. The torrent of the circulation thus furnishes the muscles with enormous stores of heat, which they are able to convert into work. Thus, while the temperature of man remains fixed at about ninety-eight degrees Fahrenheit, that of the bird reaches one hundred and nine to one hundred and eleven degrees. It exceeds, consequently, the limits beyond which our organs become unfitted for life. It has been proved that the bird consumes, in a state of repose, a great quantity of oxygen ; one would undoubtedly be astonished if it were possible to ascertain how much it absorbs in rapid flight. Let us add that, in order to endure this active combustion, the bird must be able to repair promptly the losses it suffers. Its organs of nutrition respond to this necessity. Its gizzard, hard as horn, grinds without difficulty the most resisting articles of food ; a liver of great size pours torrents of bile over the material which comes from the gizzard, and digestion is effected with surprising rapidity. So the bird cannot go hungry. It is sometimes remarked of a person who takes but little nourishment that he eats like a bird. This is a phrase to be taken with reservation, and one that should undoubtedly be removed from our vocabulary. The species which feed on living prey make a very great slaughter ; those which live on fruits and grains eat a little at a time perhaps, but it is on the condition of finding the table always spread.

III.

Muscular Contraction and Innervation.

We have just established in the case of man, and incidentally in the case of birds, the conversion of heat into work. We must again examine, a little more closely, the circumstances which accompany this phenomenon.

The muscles swell up, and become shortened, to effect the motions of the bones to which they are attached.

When, in physiological experiments, a muscle is made to contract by an artificial irritation, pinching it, for example, or communicating to it an electric shock, there result jerks and violent contractions, which do not resemble the regular motions which the will calls into action; but if a continued series of irritations be kept up, the muscle * is observed to contract in a

* Muscles are composed of fibres, which are contractile in a a high degree, and which occur under two principal forms : the smooth fibre belongs to the muscles which serve the purposes of organic life, to that silent, and as it were, unconscious life, which animates the various parts of the body; the striped fibre belongs to the muscles of the life of relation, those which produce the voluntary motions. Certain muscles, the heart, for example, present a mixed composition. There would appear to be a dif-

permanent manner. Helmholtz, employing the interrupted current of an induction coil, has shown that at least twenty-two excitations per second are needed to secure continued contraction.

The muscle thus contracted gives out a perceptible, though very deep sound. Helmholtz was able to prove that the pitch of this sound corresponded to the number of interruptions produced in the induction coil.

There is, besides, a characteristic fact that accompanies muscular contraction, one that may be regarded as its direct cause; it is a powerful absorption of oxygen. M. Matteucci proved this by comparing, in a lime-water bath, the amount of carbonic acid given out by muscles, when contracted, and when in a state of rest. The oxidation of muscles is likewise directly observed in the animal economy; it is known that the venous blood, when it comes from muscles a long time contracted, is completely deprived of oxygen, and contains a large excess of carbonic acid.

Thus there is no doubt in regard to this. What

ference in the mobility of these two kinds of fibres. The striated muscle, when irritated, contracts abruptly, and relaxes immediately; the smooth fibre acts more slowly, and in a more prolonged manner. Physiology has chiefly studied the striped muscles; it is these which possess the highest importance for us at this time, since they are the instruments of voluntary motion.

designates contraction is an increase of energy in the oxidation of the muscular tissues, a more active decomposition of the hydrocarbonated materials by the elements of the arterial blood. That chemical action thus set up throughout the extent of the muscle changes its form, that it shortens the muscle while increasing it in size, is not a matter of astonishment to us; we often see a rope swell up and become tight when it is wet, and produce in this way a considerable traction. That the heat developed in muscular tissue should be partially converted into work, is what we also regard as an ordinary and common phenomenon. M. Beclard has, moreover, made a series of ingenious experiments upon this matter. He has studied, in its calorific bearing, the same muscular contraction in the case where it produces no external work, and in the case where it does; he thus observed, during a long series of experiments, that the heat due to chemical action was diminished by just that amount which was transformed into work.

But let us not stop here; let us endeavor to go back to the origin of muscular action.

The nerves interpose to excite the action of the muscles.

The nervous system, if we regard it in its relations to motion, may be represented in the following manner. An external organ receives the sensations; a very slender tubular filament carries them to a nerve cell which perceives them; another cell, suited to direct movements, communicates by means of a new filament with the contractile apparatus that is to execute them; finally, a nerve tube acts as a bond of union between the cell which receives the impressions and the motor cell. Such, reduced to its simplest expression, is the general idea of nervous communication. The act which is propagated from one extremity to the other of the system is termed a *reflex act.* The elementary filaments, very thin and delicate, since a great number of them possess a thickness of scarcely a hundredth part of a millimeter, are bound together and intertwined in such a way as to form little cords; the cells are also grouped together at certain points, which bear the name of nervous centres. With vertebrates, with man, whom we have especially in view, the greater number of these nervous centres are united together in that long stem, which constitutes the spinal marrow. Nevertheless a certain number remain, which are scattered throughout the body; these are called nervous ganglions, and taken together they are known under the name of the great sympathetic system. A sort of hierarchy is thus established in the reflex actions; some concern the ganglions only, while others reach as far as the spinal marrow. Above this rises a still higher system. At the origin of the encephalon are found the oval masses, which preside over the respiratory movements and the contractions of the heart; next comes the cerebellum, which co-ordinates the voluntary motions, then the lobes of the brain, where the will and the intellect reside. First the ganglions, afterwards the spinal marrow, make successively a sort of selection from among the reflex acts, permitting only a certain number of them to reach the higher regions of the system where would seem to be concentrated the conscious governing power of the being. Thus may be reduced to a few general outlines the infinite complication of this so delicate network which ramifies throughout the whole extent of the body.

How is nervous action propagated?

Several years ago the works published by M. Du Bois Reymond, and several German physiologists, seemed to have solved this problem. A solution which offered itself, under such an attractive exterior, was accepted with eagerness. Innervation was an electric current. A current was transmitted along the sensitive nerve to end in the cell of sensation; a current quitted the motor cell in order to end in the organ of motion; whatever might have been the reactions effected within the cells, they assumed, henceforth, a manifest analogy with that which takes place in a battery, or other electro-motor machine.

The excitement over this explanation cooled off. Admitted at the outset, with insufficient proof, it was finally rejected by many physiologists without sufficiently good reasons. We do not find in the human body the simple conditions presented by our electrical apparatus. It is clear that a nerve cannot be entirely analogous to an insulated conducting arc, since it is itself, like everything around it, the seat of incessant reactions. One was too readily discouraged by reason of the confusion in the results yielded by experiments. To invalidate the existence of nervous currents, reasons are brought forward which do not appear to have great weight. Electric currents, it is said, are transmitted slowly in the nerves, having only a velocity of twenty-four, or even of eighteen metres in a second; they go less rapidly in the nerves than in the muscles. It is argued again that a nerve, when cut, however closely the ends may be applied to each other, becomes unfit for communication. These are matters which have nothing of a decisive character. Whatever may be said, we find ourselves still confronted with important and highly significant facts. By causing electric currents to act upon a nerve,—veritable currents produced by our machines,—we obtain the contraction of muscles; not an instantaneous contraction alone, but a continued one. Let one take the hinder parts of a frog, the two thighs attached to the lumbar nerves, and these latter to a fragment of the spinal marrow, and let a current be passed along one of the nerves, and there will follow not only a direct excitation of the corresponding limb, but also the reflex motion of the other thigh.

It seems to us that these results, well known and within the reach of common experience, furnish serious grounds for conviction. Now, if it be proved that the stream which reaches the muscles is not to be confounded with the electric current, that it is to be regarded as possessing a special character, and to be studied under a distinct name, there will yet be nothing in this circumstance that can weaken the results we present. Under cover of this declaration, we continue to speak of nervous action as of an electric current. It will be possible, if so desired, to see in this language merely a figurative representation of the phenomena; it will be sufficiently exact to justify the results we wish to illustrate.

Thus the nerve excites the muscle. Does this mean that the nerve possesses in itself all the energy which is developed in the muscle? No; since the muscle gets this force directly by its own oxidation. The nerve merely excites chemical action; it only sets going a piece of machinery. It is thus that a spark produces the explosion of a gaseous mixture; it is thus that a match effects the lighting of a fire; it is thus, by turning a stop-cock, that we let flow away all the water accumulated in a reservoir.

One is naturally led to think that the work of the nerve is exceedingly small compared to that of the muscle. M. Matteucci has proved this a fact by direct experiment. He hung a weight to the principal muscle of the leg of a frog, and sent an electric current through the nerve attached to this muscle. The contraction of the muscle raised the weight, and it was easy to estimate the effort in foot-pounds. Likewise, by a simple calculation, might be estimated the combustion of the zinc produced in the battery during the very brief period of excitation. M. Matteucci thus found the work done by the muscle to be at least twenty-seven thousand times greater than the chemical or calorific effect of the nervous excitation.

Let us go back yet farther, and approach the origin of the motion. Small as the work of the nerve may be, how is it accomplished? To give rise to a current in a nerve, it is enough that a circuit be formed somewhere, either within or on the outside of the nerve cell, and this action itself is only a very slight portion of the action which the current can produce. As to the mode in which such a circuit is formed, nothing definite can be known. If it be a question of voluntary motion, we say that the will intervenes.

But two things are to be noted.

First, the mechanical action attributed to the will becomes, from the preceding considerations, gradually reduced to one so extremely minute that it seems to disappear altogether.

Let us add, in the second place, that the will does not create this work, however imperceptible it may be. It can only be conceived of as a special agent of transformation in motions infinitely small. The voluntary act, — and for a still stronger reason the purely reflex act, — to whatever degree of tenuity it be reduced, does not proceed without a subtle modification of the tissues in which it is effected, without I know not what sort of delicate transformation of molecular motions.

In ascending from muscular action to nervous action, properly so called, and to the play of the will, we have reached the limit where physical phenomena give place to moral, and beyond this we have not to pass.

Within the limits we have now reached we have

been able to show how the principles to which we have been led by the study of the inorganic world are verified in the case of living beings. Our conception of the physical universe would have been too incomplete had we been obliged to curtail from it all that pertains to life.

We may now, without leaving behind us so formidable a gap, resume the synthesis we have undertaken, and endeavor to give it its final shape.

CHAPTER VII.

CONCLUSION.

CUVIER said, in his History of the Progress of the Natural Sciences, "Once having quit the phenomena of shock, we no longer possess any clear idea of the relations of cause and effect. Everything is reduced to the collecting of particular facts, and the searching out of the general propositions which embrace the greatest number of them. It is in this that all physical theories consist, and, to whatever degree of generality each of them may have been reduced, they will still be far from a conformity with the laws of shock, which alone could change them into real explanations."

It cannot be said that physicists have already fulfilled the programme marked out by these words. And yet, if we cast a look behind over the road we have traversed, and include in a general view all the facts we have mentioned, we shall feel more and more established in this idea, that all physical phenomena consist in the exchange and transformation of material motions.

Will it be said that our examination has not always been rigid enough? that we have sometimes asserted, when we ought to have expressed a doubt? that we have not always laid sufficient stress upon the reservations we were led to make? We shall not try to shield ourselves from this censure, feeling too well that we have deserved it. It would have been better, perhaps, to have left more points in obscurity, and to have limited ourselves to certain facts. May we be pardoned for some suggestions too conjectural? The results acquired are considerable, and a few rash suppositions cannot compromise them.

These acquired results we could at need clothe with the authority of an eminent man of science. M. De Senarmont, during the last year of the lectures which he delivered with so much brilliancy at the *Ecole Polytechnique*, and which death came so early to interrupt, thus summed up his views upon the progress of the physical sciences: "Even lately each group of facts acknowledged a special principle. Motion and rest resulted from forces, ill-defined enough specifically, but which it was agreed to term mechanical; the phenomena of heat, light, electricity, badly enough defined themselves, were produced by so many particular agents, fluids endowed with special activities. A more searching examination has enabled us to discover that this notion of a variety of specific and heterogeneous agents has at bottom only a solitary and unique basis; this is, that the perception of these various kinds of phenomena is in general effected through the different organs, and that in addressing themselves more particularly to each of our senses, they necessarily excite particular sensations. The apparent heterogeneity would then be less in the nature itself of the physical agent than in the functions of the physiological instrument which shapes the sensations; so that in falsely attributing the dissimilarities in the effect to the cause, we should have in reality classified the intermediate phenomena through which we have a knowledge of the modifications of matter, rather than the essence itself of these modifications. . . . All physical phenomena, whatever their nature, appear at bottom to be only manifestations of one and the same primordial agent. . . . This general result of all modern discoveries can no longer be ignored, however impossible it may still be to put into definite shape their laws and their conditional details."

Such was the language of M. De Senarmont in a course of academical instruction, in which no room was allowed for any unsafe doctrine.

We are not bound to a like reserve. So we have more explicitly stated the system which seems to sum up the labors, and to express the general sentiment of contemporary physical science. Ether in a state of motion fills all space. The ethereal atoms, by their aggregation, form molecules ; these last, bodies. Between these atoms, these molecules, these bodies, interchanges of motion take place, constituting what we term heat, light, electricity, gravity, chemical affinity. These interchanges depend upon the masses and velocities which are concerned. The conception of the physical universe is contained entirely in these principles. Hitherto we have been able to get but a very small number of the facts which this statement of principles embraces, because we are unacquainted, nearly always, both with the absolute and the relative values of the masses and velocities which govern the communication of motions. Practically we are satisfied with saying that heat, light, electricity, gravity, affinity, are transformed into each other according to the fixed relations of equivalence, and we assign them a common measure, that of mechanical work.

Left thus on the outside of phenomena, we have but a vague notion of the circumstances that accompany and determine transformations. There are, doubtless, motions to which we are unable to give a name, and which we are not qualified to perceive, although they play their part in nature.

Amid the variety and number of motions that would seem possible, why are some produced and not others?

Is there among motions a sort of natural selection?

We should possess the key of the transformations taking place under our eyes if we could attain to that measurement of the masses and velocities which, until now, eludes us. In a steam engine, for example, the agitation which reigns in the heat of the furnace is communicated to the tubes of the boiler, and from these to the water itself; the molecules of vaporized water each expend a little of their living force upon the piston, which moves under these accumulated efforts, and sets in motion the shaft of the engine; but we perceive this series of changes only through a veil. When a motion of a certain kind is replaced by another of a different kind, the reason for this exchange usually escapes us; and it is because of this ignorance that we have recourse to the idea of force; we say that a force is exhibited, and produces such an effect, because we are unable to grasp the anterior motions from which this effect results.

The notion of physical force ought then to disappear if the elements of molecular mechanics were known. In the present state of our knowledge we must, indeed, preserve it; but we must also be on our guard against the errors it may entail. Let us call force every cause of motion, if it be desired; but let us not forget that this word most generally represents only a provisional and conditional cause. The dread of a vacuum has been a force in its time, that is, of a vacuum to the extent of thirty-two feet.

If we recur with persistence to this consideration, it is because it seems to us to be of capital importance, and we could not devote too great an effort towards its illustration. It is the knotty point in the system we have unfolded. And yet, among the physicists even who have entered into the current of new ideas, there is a school which persists in giving to the physical forces an unaccountable individual existence.

M. Hirn, whose name in France is connected with the determination of the mechanical equivalent of heat; M. Hirn, whom we mention when we wish to place a French name by the side of those of MM. Joule and Mayer; M. Hirn does not hesitate to regard the physical forces as the constituent elements of the universe. Under the title of *intermediate principles*, he makes of them half transcendental essences, which fill all space, and which have the property of conferring motion upon matter. He even makes an enumeration of these principles, and finds four of them gravity-force, light-force, heat-force, and electric-force.

What! does matter here and there quit its state of rest, and do new motions spring up at the will of these forces? This is not precisely what M. Hirn means; he knows too well it would be in contradiction with the facts. Here is the theory he conceives. For him each force is everywhere diffused. At the instant the intensity of one increases in such a way as to produce a motion, the intensity of another force diminishes in a corresponding ratio. Now this diminution of intensity in the second force itself corresponds to a diminution of motion in matter. It is, evidently, a sort of pre-established harmony. Doubtless we have but to suppress these artificial intermediate agents, to find ourselves face to face with motions themselves, and the return is thus easy, when it is desired, from M. Hirn's stand-point to that we just now occupied. Why introduce, hereafter, between two motions which beget each other, two semi-transcendental essences? Why have recourse to these intermediate principles? Why this mythology, this Olympus of forces?

Why? It is not very hard to give the reason, nor will it be useless to do so.

These arbitrary conceptions are inspired in M. Hirn by the disturbing influence of an easily alarmed spiritualism. M. Hirn becomes distrustful when he sees a doctrine which every day explains by the motions of matter an ever-increasing number of facts. He dreads the encroachment. He fears lest it may come to reach the human soul; lest it reduce to pure motions the phenomena of will and of thought. It is for the purpose of arresting it in its progress that he has recourse to gravity, heat, light, and electric forces. These intermediate principles are the bulwarks he raises to defend the soul-principle. Strange bulwarks in truth, and far from capable of such a defence! Must it be repeated, moreover, that the problems of the soul are in no way concerned in the theories against which M. Hirn endeavors to fortify himself? In the midst of material transformations, causes active in themselves may interfere, and we have pointed out examples of them in indicating the nature and limits of this interference. This is sufficient to leave the field free for all the solutions of metaphysics.

Having shown how our hypothesis banishes the fallacious entities with which physical science may be encumbered, is there need of defending the theory itself from the extravagant deductions that might be drawn from it? Is there need of indicating the point of view from which an entirely healthy conception of it is to be obtained?

Is admitting a scientific hypothesis equivalent to believing one's self in possession of the realities of things? This would be too easily forgetting so many systems that have buried each other in their ruins. It would be forgetting too easily that the physical philosopher, lost in the infinity of time and space, seizes only apparent relations, and does not even arrive at a conception of the absolute! What, then, is it to group together into an hypothesis all our ideas concerning nature? It is to afford us the means of illustrating the things we learn by comparing them with each other, of establishing fruitful relationships between facts, and so cause springs of discovery to burst forth.

What is of importance, correctly speaking, in such an hypothesis, is not the picture it gives of nature, but the plan it traces out for the explorer of physical science.

In this connection, the system we have exhibited is admirably summed up in a solitary principle. It evolves a luminous criterion, whose efficiency has already been revealed in scientific researches.

This precious symbol has a name in the language of mechanics; but before pronouncing it, let us hasten to recall to mind what we have already said regarding the difficulty one encounters of expressing new ideas with old words. By a cruel irony of circumstances, we are about to fall in with a word we would like to have escaped from at this time by reason of the ambiguity it contains. Never have we more keenly felt the need of employing a new expression, and if we refrain from so doing, it is for the reason that our declaration in this regard will doubtless stand us in stead of a neologism. We conceive the presence in the universe of an unchanging supply of material atoms endowed with rapid motion, and grouping themselves into systems to form molecules and bodies. Each of the atoms and systems possesses, in proportion to its mass and velocity, what we have hitherto termed a *living force,* — what we may now call, if we desire to avoid this ambiguous term, an *energy*, without gaining much by the change. Such are the expressions against which we have desired to take precautions by a preliminary statement. We do not employ a new word; but we have said enough to show that under these usual designations, we are to see absolutely only masses in motion. To say that energy changes its locality, is merely to say that masses act upon each other in mutually modifying their velocity.

Energy thus passes without limitation from one system to another, thus giving origin to the variety of natural phenomena. Sometimes it shows itself in a series of changes, in which its successive efforts may be followed; then it is said to preserve the *active* form. Sometimes it hides itself in order to maintain, for a longer or shorter period, an equilibrium, whose rupture will regenerate it; it is then said to pass into the *potential* state. Active energy and potential energy vary incessantly in their relative proportion, but their sum remains constant.

Such is the principle usually designated under the name of conservation of energy.

Doubtless, in order to verify entirely this constancy of energy, it would be necessary to include the whole universe. Energy may increase at certain periods, in certain regions of space, and decrease in different regions, though the ether would appear to be a sort of regulator of this universal activity. Do the changes which unceasingly take place between our terrestrial globe and the sidereal medium become interpreted to us by a loss, by a gain, by a periodic oscillation about a mean condition? How does our solar system comport itself with reference to other systems? Such are the vast problems to which the notion of universal energy finds its application.

We do not mean that the principle of the conservation of energy cannot be verified in the immediate connection of ordinary phenomena. It establishes a definite bond of union between all the facts which surround us. The physical philosopher knows that motions can pass from visible masses to invisible ones, without ceasing to obey a law whose purport he knows. If he is not always fortunate enough to gather the facts into complete cycles, where effects and causes are linked into a chain whose ends meet, at least he is no longer forced to regard phenomena as isolated appearances. In the case of each one he is able to ascend to its sources, or descend to its consequences. He may fail in the application of his method, he may represent to himself in a false light, this or that family of facts, but the principle itself, by virtue of which he seeks a fundamental unity beneath the infinite diversity of appearances, is to him the most precious and the best assured conquest of contemporary science.

INDEX.

A.

B.

C.

D.

E.

F.

G.

H.

I.

J.

K.

L.

M.

N.

O.

P.

Q.

R.

S.

T.

U.

V.

W.

X.

Y.

ERRATA.

Page 18, line 15, *for* organic *read* inorganic.
" 41, last line, *supply the word* our *before* thought.
" 47, line 2, *for* Anaximine *read* Anaximines.
" 48, line 18, *for* advantages *read* advantage.
" 80, line 18, *after word* component *insert* of.
" 81, line 21, *for* good *read* goal.
" 195, line 19, *for* agreement *read* argument.

The notes at the bottom of pages 78, 80, 86, 118, 120, 164, are by the translator.

The note at the bottom of page 188 is by the author, as far as "The reader will observe," etc.

IMPORTED BOOKS.

CHAMBERS' ENCYCLOPÆDIA. A Dictionary of Universal Knowledge, in which the facts of Science, Philosophy, History, and even matters of Familiar Conversation, are given in a manner adapted for easy consultation. Profusely Illustrated by Wood Engravings and Maps. This comprehensive and elaborate work is now complete in 10 volumes, 8vo.

Cloth, including Maps, ... $45.00
Half calf, marble edges, including Maps, 65.00
" Morocco, extra, ... 70.00

MORAL EMBLEMS. With Aphorisms, Adages and Proverbs, of all Ages and Nations. From J. Cats and Robert Fairlie. With Illustrations by John Leighton, F. S. A. Quarto.
Full gilt, cloth, extra, ... $12.00

MRS. ANNA JAMESON'S WORKS ON SACRED ART, as follows: —

Sacred and Legendary Art. Illustrated with 19 Etchings and 187 Wood Engravings. 2 vols.

Legends of the Monastic Orders. Illustrated with 11 Etchings and 88 Wood Engravings.

Legends of the Madonna. Illustrated with 27 Etchings and 165 Wood Engravings.

The History of Our Lord, as exemplified in Works of Art. Illustrated with 31 Etchings and 281 Wood Engravings. 2 vols.

In all 6 vols., demy 4to.
Cloth, ... $40.00
Half calf, gilt tops, ... 65.00
" Morocco, ... 67.50
Tree calf, ... 75.00
Morocco, extra, gilt edges, ... 100.00

www.ingramcontent.com/pod-product-compliance
Lightning Source LLC
Chambersburg PA
CBHW021518210326
41599CB00012B/1305
* 9 7 8 3 7 4 4 6 9 0 3 7 9 *